KT-460-315

PLANET EARTH

PLANET EARTH

Derek York
Professor of Physics
University of Toronto

McGRAW-HILL BOOK COMPANY

*New York St. Louis San Francisco Aukland Düsseldorf
Johannesburg Kuala Lumpur London Mexico Montreal
New Delhi Panama Paris São Paulo Singapore
Sydney Tokyo Toronto*

This book was set in Times Roman by Black Dot, Inc.
The editors were Robert H. Summersgill and James W. Bradley;
the cover was designed by J. E. O'Connor;
the production supervisor was Judi Frey.
The drawings were done by Eric G. Hieber Associates Inc.
The Murray Printing Company was printer and binder.

PLANET EARTH

Copyright © 1975 by McGraw-Hill, Inc. All rights reserved. Printed in the
United States of America. No part of this publication may be reproduced,
stored in a retrieval system, or transmitted, in any form or by any means,
electronic, mechanical, photocopying, recording, or otherwise, without the prior
written permission of the publisher.

1234567890 MUMU 798765

Library of Congress Cataloging in Publication Data

York, Derek.
    Planet earth.

    Includes index.
    1. Earth sciences. I. Title.
QE28.Y62   551.1   74-20874
ISBN 0-07-072290-0

To
Lydia, Linky, Alice, John
and
J. A. Holden

Contents

Preface

During the past 20 years, our understanding of the solid earth has been revolutionized. As a result we now know far more precisely than ever before its age, its shape, its internal structure, and the detailed behaviour of its magnetic field. The crowning discovery, however, has been that the continents have undoubtedly been drifting about the earth's surface for hundreds of millions of years at speeds of a few centimetres per year. They have been doing this via an ocean-floor-spreading mechanism which sees brand new ocean floor created along great rifts while old sea beds are returned to the earth's interior for recycling. In this book I have tried to present a clear and simple account of many of these developments. To make the book meaningful to many, I have avoided mathematics but have still tried to give the details of the more important aspects involved. I have also adopted the historical approach to the subject. Earth science has a fascinating history, and I have referred to it throughout, though most obviously in the final chapter.

I hope "Planet Earth" will inspire the reader to delve further into a study of the earth, for what I have written is not complete. Clearly in a book this size not everything known can be recounted, but more importantly not everything about the earth is known. The reader should keep this in mind when going over some of the triumphs documented ahead.

I have tried to explain various concepts as they arose so that the book could be used and understood without continual reference by the reader to other sources. As such, it should be comprehensible to both nonscientists and scientists, to students at the university level or in their final year at high school, and to anyone wanting an introduction to the large-scale features of the "solid earth."

In Chapter 6 a key concept is called the "Vine-Matthews hypothesis." There is evidence that the two Canadians L. W. Morley and A. Larochelle should be regarded as co-inventors of this idea, and it may be that it will come to be labelled the "Vine-Matthews–Morley–Larochelle hypothesis."

Finally, I thank my wife for encouraging me continually during the writing of this manuscript, and I express my deep appreciation to James W. Bradley and Robert H. Summersgill of McGraw-Hill, with whom it has been a pleasure to work.

Derek York

PLANET EARTH

Chapter 1

Genesis

No discussion of the earth and its evolution would be complete without some mention of the universe as a whole, and so we open our story by looking out from the earth and surveying our cosmic environment. While there are quite literally trillions of objects around us in space, our survey is simplified by the fact that the various bodies are not distributed in entirely random fashion but rather are gathered into various family groupings. The most immediately obvious of these is the solar system. This comprises mainly a central star, our sun, around which revolve nine planets: Mercury, Venus, Earth, Mars, Jupiter, Saturn, Uranus, Neptune, and Pluto, in order of distance from the sun. Looking beyond our solar system on a clear night, we see, with the naked eye, over 2,000 stars; with a pair of binoculars this is increased to about 100,000 stars, while over 1 billion stars may be detected with the gigantic Mt. Palomar telescope. Overhead, and stretching across the sky, we see a hazy band of light known as the *Milky Way*. When Galileo (1564–1642) turned one of the earliest telescopes towards this band, he observed that it consisted of many faint stars and that it provided a clue to the existence of the next family grouping we shall consider, our galaxy. The sun and its planetary system, and about 100 billion (usually written as 10^{11}, where 11 refers to the number of zeros occurring after the 1 when the number is written in full) other stars, are assembled into a disc-like structure which we call a *galaxy* and illustrate in Fig. 1-1. The disc has a spiral structure and is slowly rotating rather like an enormous Catherine wheel, although the internal motions are by no means uniform. One full galactic rotation consumes approximately 240 million years, which is sometimes referred to as the *cosmic year*. The solar system, which participates in the galactic rotation, is situated two-thirds of the way towards the periphery of the disc, as shown in Fig. 1-1. When looking in the direction of the Milky Way, we are looking along the plane of the galactic disc and consequently see a much greater density of stars than when we are looking out from the disc. Telescopic observation further reveals that there are many and varied galaxies in the sky (Fig.

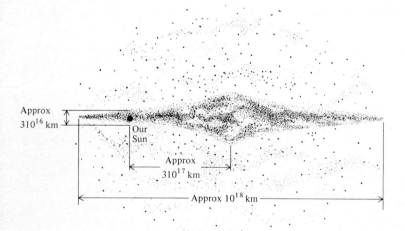

Approx 310^{16} km

Our Sun

Approx 310^{17} km

Approx 10^{18} km

Figure 1-1   The position of the solar system in our rotating galaxy.

1-2). Approximately 10 billion (10^{10}) galaxies are observable, and it is not known how many more remain undetected.

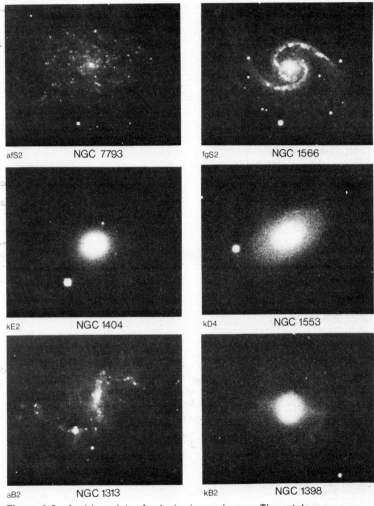

Figure 1-2   A wide variety of galaxies is now known. The catalogue number and form (S = spiral; E = elliptical; D = dust-less; B = barred spiral) are given. (*Courtesy R. F. Garrison, David Dunlap Observatory, University of Toronto*.)

Aside from the assemblies we have already mentioned, it is notable that the stars within galaxies frequently form clusters and that there also occur clouds of dust and gas. The dust clouds, in fact, obscure much of the Milky Way as far as visual observation is concerned, and the essential shape of our galaxy has been determined by radioastronomy, since the longer radio waves are essentially unaffected by dust particles.

THE EVOLUTION OF THE UNIVERSE

All theories of the origin and development of the observable universe are profoundly influenced by one fact of experimental observation, the *red shift*. Working at the Lowell Observatory in the United States, W. M. Slipher had by 1914 determined the spectra of 14 galaxies. The wavelengths of the light from all but two of these were displaced towards the red end of the spectrum. Two showed blue displacements. Now the most reasonable explanation of these observations is based on the Doppler effect. It is a well-known fact, verifiable by experiments conducted in the laboratory, that such a reddening of light is observed when the light source is moving away from us: the greater the velocity of separation, the greater the red shift seen. Conversely a blue shift is produced when the source and the observer approach one another. Reasoning along these lines, therefore, Slipher calculated that most of the galaxies he had studied were receding with velocities in range of 100 to 200 miles/second. Further measurements showed the vast majority of galaxies to be moving away from us with similarly enormous speeds. Finally, in 1929, the American astronomer Edwin Hubble, working at the Mount Wilson Observatory, made the profound discovery that the more distant a galaxy the greater is its red shift and, therefore, its presumed recessional velocity. Furthermore, there is an approximately linear relation between the velocity of recession of a galaxy and its distance (Fig. 1-3).

Big-Bang Theory

This linear velocity-distance relation strongly suggests that at some time in the past all the galaxies were concentrated near ours and were blown to their present positions by an enormous

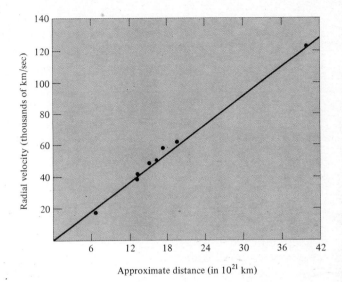

Figure 1-3 Hubble's linear relation between the velocity of recession of a galaxy and its distance.

explosion. After such an explosion the faster moving galaxies would naturally be farthest from the centre. It would be wrong, however, to assume that this implies that the centre of such an explosion would have to have been located in space roughly where our galaxy is now. Simple relative-velocity considerations show that after an explosion an observer situated on *any* flying fragment would see all the surrounding fragments receding from himself with a linear velocity-distance relationship, just as though the explosion had been centred on his current position. Such considerations quite naturally led to the theory that all the matter of the universe was at one time concentrated in a very small volume which subsequently blew apart with great violence. The Belgian mathematician Lemaître called this initial agglomeration the *primeval atom*. The rapidly expanding material must have been in the gaseous state (apart from the radiation), and it is presumed that density fluctuations in this gas enabled gravitational forces to cause local condensations of the gas in such a manner as to yield the stars and galactic systems.

It is obviously tempting, on this model, to extrapolate the

expansion backwards in time to determine when the state of extreme concentration occurred. If we assume that gravity has not been acting steadily to slow down the expansion, we can calculate the time of the initial explosion in an elementary fashion by dividing the distance of a galaxy from us by its recessional speed. When this is done for various galaxies, it appears that the initial explosion occurred approximately 10^{10} years ago. That such an estimate is indeed very approximate may be seen from the fact that, as recently as 1965, Gamow maintained that the *big-bang* expansion of the universe began only 5×10^9 years ago.

Steady-State Theory

In 1948, the applied mathematicians Bondi and Gold and the astronomer Hoyle proposed an entirely different picture of the evolution of the universe, yet one which contained the red-shift observation as a fundamental aspect. It was accepted that the red shift implied the recession of all the galaxies from one another, but the important principle adopted was that the universe remained basically constant in appearance throughout both space and time. If a constant density of matter is to be maintained at the same time the galaxies are all receding from one another, one obviously has to create new matter out of which new galaxies can condense. Bondi, Gold, and Hoyle therefore proposed that there was indeed a continuous creation of matter, one hydrogen atom per year supposedly appearing spontaneously per unit of volume equal to that of St. Paul's Cathedral throughout the universe. While this ad hoc postulation of spontaneous creation is distasteful to many, it is also true that such a universe with no beginning and no end is appealing to others.

There are in existence two pieces of evidence which favour the big-bang theory over that of the steady state. Both are drawn from the field of radioastronomy. In 1961, Ryle and his coworkers published the results of a radioastronomical survey in which they examined the radio emissions from a wide distribution of galaxies. They observed that there seemed to be more powerful galactic radio sources at great distances from us than there are in our immediate neighbourhood. Since the radio waves that we receive now from these distant galaxies were emitted several thousand million years ago, in looking to these distant galaxies we are looking backwards in time. The results of Ryle and his coworkers' survey therefore imply that there were more powerful radio-wave-emitting galaxies several thousand million years ago than in much more recent times. Clearly this favours a model of the universe (such as the big-bang theory) which allows evolution with time over the steady-state model.

The second piece of radioastronomical evidence supporting the big-bang theory appeared in an interesting and roundabout fashion. In 1948, Gamow proposed that in the beginning phase of the explosive expansion of the universe radiation was dominant over material objects and that the radiation was blackbody radiation. The latter is radiation which would be found in a completely closed container when the radiation is in perfect equilibrium with the walls. Such radiation has a continuous spectrum similar to that shown in Fig. 1-4. As the temperature of the container is raised, the shape of the spectrum remains little altered, but the pattern is shifted to shorter wavelengths. (The radiation emitted by a hot poker is essentially blackbody radiation, and the change in colour of a poker as it is heated through red to white heat corresponds to this shift of the peak in the spectral distribution to shorter wavelengths.) Such blackbody radiation present at the beginning shared in the universal expansion. As the universe expanded, its blackbody-radiation temperature dropped. Gamow calculated that when the universe had been expanding for one hour its temperature was 250 million °C. After 200,000 years, the temperature was about 6000°C, while on its "250 millionth birthday" it was about 100°C below the freezing point of water ($-100°C$). At the present time the temperature of the universe would be within a few tens of degrees of the *absolute zero of temperature* if the expansion had proceeded for 5 to 10 billion years. (The absolute zero of temperature is 273.15°C below the freezing point of water and, according to the third law of thermodynamics, is the lower limit of temperature. That is, in a finite length of time, no body can be cooled to this temperature. The lowest temperature actually achieved in a laboratory was 0.0000000006°C above absolute zero by physicists in Saclay, France, in 1969.) At such a temperature (a few tens of degrees

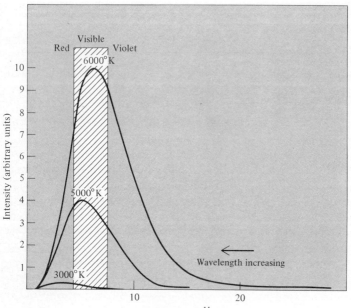

Figure 1-4 The spectrum of blackbody radiation. At higher temperatures the peak in the distribution is at shorter wavelengths. (°K = °C + 273.15.)

above absolute zero) the blackbody radiation is concentrated in the microwave range (10^9 to 10^{12} Hz, i.e., cycles/second) and would be detectable by radioastronomical methods. Throughout the 1950s, however, no attempts were made to detect such background radiation. Then in 1965 at the Bell Telephone laboratories in Holmdel, New Jersey, A. A. Penzias and R. W. Wilson found some distinct residual radio noise in their receiver despite elaborate attempts to minimize all the usual well-known noise contributions. A group of physicists collaborating with R. H. Dicke at Princeton University suggested (unaware of Gamow's work) that this residual noise in fact represented the background blackbody radiation of the universe. Penzias and Wilson carried out their radiation-intensity measurement at a wavelength of 7.35 cm, while about 20 later intensity measurements have been performed at both greater and lesser wavelengths. These results fall on or near a blackbody-radiation curve equivalent to a temperature of about 3°C above absolute zero (see Fig. 1-5), which is a very striking result in keeping with the big-bang theory. However, it is seen in Fig. 1-5 that the reliable data fall on just one flank of the blackbody hump. Obviously the results would be much more convincing if precise data lying on the shorter wavelength side of the curve were obtained. Unfortunately there is a major problem involved in tracing out the complete blackbody curve. Radio observations may be made only through a "window" stretching from a few centimetres to several metres wavelength. Incoming radio waves outside this range are absorbed by either the atmosphere or the ionosphere. As may be seen from Fig. 1-5, therefore, the wavelengths corresponding to

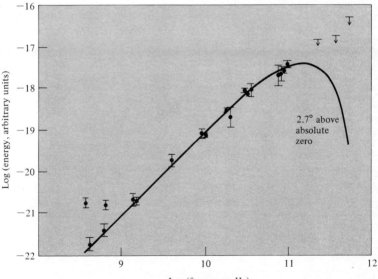

Figure 1-5 The data so far obtained on the background radiation lie near a blackbody curve corresponding to a temperature of 2.7°C above absolute zero. The less accurate points to the right of the peak in the curve were obtained by indirect means. (*After M. J. Rees.*)

the short-wavelength side of the blackbody curve are not detectable at the earth's surface. The data points at 2.6 mm and shorter wavelengths were accordingly obtained by indirect methods, and the full delineation of the universal blackbody curve will probably have to await more precise measurements made from satellites or rockets which are outside the befogging action of the earth's atmosphere. The potential of this approach to the explanation of the origin of the universe is clearly enormous. If the background radiation is definitely shown to exist, some kind of explosive beginning to the universal expansion would seem definitely to be called for and the steady-state cosmology would be untenable.

ORIGIN OF THE SOLAR SYSTEM

In turning from the universe at large to the origin of the solar system in particular, we are still confronted with a fundamentally unsolved problem. Any theory of solar system formation must explain the following features:

1 The sun constitutes about *99.9 percent* of the mass of the solar system but has only *2 percent* of the total angular momentum. By this we mean that the planets are moving around the sun rather quickly, considering the rate at which the sun is spinning about its own axis.

2 The planetary orbits are nearly in one plane, are nearly circular in shape, and are all described in the same direction.

3 The distances between the sun and the various planets are surprisingly well described by Bode's law. Bode, a German astronomer, noticed that if one wrote down the numerical sequence 0, 3, 6, 12, 24, 48, etc., added 4 to each number, then divided the results by 10, one was left with a sequence of numbers which represented with striking accuracy the distances of the planets from the sun in astronomical units (one astronomical unit is the mean distance of the earth from the sun). The agreement between Bode's predictions and the observed mean planetary distances is shown in Table 1-1.

4 The mean densities of the *terrestrial* planets, Mercury, Venus, Earth, and Mars, are 3.8, 4.86, 5.52, and 3.33 respectively, whereas the *outer planets*, Jupiter, Saturn, Uranus, and Neptune, are much lighter, with densities 1.34, 0.71, 1.27, and 1.58 respectively.

5 Hydrogen and helium constitute about *99 percent* of the sun's mass, with the other known 90 chemical elements contributing a mere *1 percent*. The terrestrial planets on the other hand contain large quantities of such elements as oxygen, silicon, and iron. Enormous amounts of hydrogen and helium are again thought to be present in the outer planets.

Numerous theories have been proposed to explain these and other features of the solar system, but none has been worked out in satisfactory detail. Currently the most popular is a modification by Von Weiszacker and Kuiper of the eighteenth-century nebular hypothesis of Kant and Laplace. According to this theory, the solar system is supposed to have originated as a rarefied cloud of gas and dust particles in slow rotation. The chemical composition was approximately that of the present sun. Under the influence of its own gravitational field the cloud began to collapse, with a consequent increase in the rotation rate to preserve angular momentum, just as a spinning skater's outflung

Table 1-1   Bode's Law.* (After Payne-Gaposchkin)

Planet	Series (+4)	Sum/10	Distance from sun (astronomical units)
Mercury	4 + 0	0.4	0.39
Venus	4 + 3	0.7	0.72
Earth	4 + 6	1.0	1.00
Mars	4 + 12	1.6	1.52
Asteroids	4 + 24	2.8	2.68†
Jupiter	4 + 48	5.2	5.20
Saturn	4 + 96	10.0	9.55
Uranus	4 + 192	19.6	19.19
Neptune	4 + 384	38.8	30.07
Pluto	4 + 768	77.2	39.52

*If Bode's law applies to all the planets, the corresponding numbers in the last two columns should be identical. Neptune and Pluto are anomalous.

†Average for asteroids.

arms when brought down to the sides quicken the spin. As contraction proceeded, the cloud flattened into a disc of turbulent particles. As most of the material condensed into the centre of the disc to form a protosun, local eddies at various parts of the disc enabled small condensations of particles to occur, resulting in the formation of protoplanets. The dust particles and the heavier elements like iron settled towards the centre of the protoplanets, and an extensive envelope of hydrogen and helium was formed. For tens of millions of years the contraction of the protoplanets and the sun continued, and in some uncertain way (perhaps involving magnetic fields) angular momentum was transferred to the protoplanets. After perhaps 100 million years the gravitational potential energy released by the contracting sun heated the sun sufficiently for the fusion of hydrogen into helium to have begun. By this time the sun was much as it is now, and the radiation streaming from it exerted enough pressure to blow away the hydrogen-helium protoplanetary atmospheres of the terrestrial planets. The outer planets were less affected, retaining considerable volumes of hydrogen and helium. A somewhat similar process on a miniature scale supposedly produced the various satellites circling some of the planets.

A detailed quantitative analysis of the above outline of solar system formation has yet to be completed, and currently one can merely say that this model allows an explanation in a qualitative manner of the features 1 to 5 of the solar system mentioned earlier.

We come down to earth now, and in succeeding chapters we will examine what is known of our planet's internal structure, its shape and gravitational field, its age, its magnetic field, and finally the evolution of its surface features.

Chapter 2

Seismology
and the Structure
of the Earth

EARTHQUAKES

For about three-quarters of a century people have used earthquakes as a source of information about the earth's structure—in
the earlier millenia they merely feared them. That this fear was
justified is well illustrated by the following account of a notorious
earthquake.

At 9:40 A.M. on All Saints' Day (November 1), 1755, the
Portuguese capital city of Lisbon was struck by the first of three
major earthquake shocks. Thousands of worshippers were killed
in devastated churches and other buildings. A second shock
arrived about 40 min later, and the third enormous blow fell
around noon. Roughly coincident with this third shock came a
tidal wave with crests about 30 ft high, sweeping away many
hundreds of people. It is usually estimated that about 30,000
people were killed in this disaster in Lisbon alone. Significant
effects from the earthquake were reported for about 3 percent of

the earth's surface. Loch Lomond, 1,200 miles from Lisbon, was set into seiche motions (sloshing movements of the water backwards and forwards in the basin) for about $1^1/_2$ hours, with water levels rising and falling about 2 ft. Lisbon had long experienced an intense seismic history, having suffered major shocks in 1009, 1344, and 1531. Following the 1755 catastrophe there was much earthquake activity for about a century, following which has come relative calm. If the history of the past 1,000 years is anything to go by, then Lisbon may well be revisited by a massive earthquake in the next 100 years. Table 2-1 and Fig. 2-1 summarize the devastation caused by more-recent earthquakes in various parts of the world.

Tidal Waves, or Tsunamis

A frequent deadly companion of the earthquake is the tidal wave, or, to give it the name more frequently used by scientists, the *tsunami*. These enormous waves are produced by earthquakes causing sudden depressions or elevations of portions of the ocean floor or by earthquake-generated submarine landslides. More local ones are originated by earthquakes causing large chunks of rock or glaciers to be dumped suddenly into the sea. When the

Table 2-1   Some Destructive Earthquakes. (After Bath)

Date	Magnitude M	Location	Number of people killed
1906 : Apr. 18	8.3	California: San Francisco	700
1906 : Aug. 17	8.6	Chile: Santiago, Valparaiso	20,000
1908 : Dec. 28	7.5	Italy: Messina, Reggio	83,000
1920 : Dec. 16	8.6	China: Kansu, Shansi	100,000
1923 : Sep. 1	8.3	Japan: Tokyo, Yokohama	99,330
1935 : May 30	7.5	Pakistan: Quetta	30,000
1949 : Aug. 5	6.8	Ecuador: Ambato	6,000
1960 : Feb. 29	5.8	Morocco: Agadir	10,000–15,000
1963 : July 26	6.0	Yugoslavia: Skopje	1,100
1964 : Mar. 28	8.5	Alaska: Anchorage, Seward	178

ocean floor is suddenly raised or lowered, the sea is set into oscillation as it tries to find its new level. These oscillations have wavelengths of over 100 miles and, in deep water, have amplitudes of only a few feet. As a consequence they are not noticed by sailors out at sea. Because of their enormous wavelength, however, they belong to the wave category known as *shallow-water waves,* even in the deepest oceans. Consequently they travel with a speed given by the expression $v = \sqrt{gd}$, where $g =$ acceleration of gravity, and $d =$ water depth. Thus, in the Pacific Ocean, with its average depth of about 15,000 ft, the average velocity is about 470 miles/hour. When these waves reach shallow coastal waters, they rapidly build up to an imposing wave front. Wave heights of over 100 ft have been authentically (and tragically) recorded. Usually the first obvious sign of arrival of such a train of waves is a strong recession of the water along a beach, somewhat akin to a very rapid going out of the tide. This is soon followed by the arrival of a series of mountainous waves.

The Pacific Ocean is particularly prone to these waves. On May 22, 1960, the coast of Chile was struck by one of the largest earthquakes on record (magnitude 8.5). About 4,000 people were killed. In addition, motion of the sea floor triggered a tsunami which spread rapidly across the Pacific at about 500 miles/hour. The waves reached Hilo, Hawaii, at midnight the same day, causing enormous damage. Wave heights were about 15 ft, and the tide gauge was put out of action. Many city blocks were devastated, and total damage was estimated at about $20 million. The underwater topography off Hilo seems to focus tsunamis on the town, and undoubtedly it will suffer repeated devastations. The loss of human life to these waves, however, has been reduced in recent years with the development of a radio warning system which alerts all susceptible areas in the Pacific following a large submarine earthquake.

Intensities and Magnitudes of Earthquakes

Table 2-2 shows the modern modification of the intensity scale of the Italian scientist Mercalli. Estimates of the *intensity,* or destructive power, of an earthquake are presented in ordered

Figure 2-1 Damage caused by various earthquakes. (*a*) Freeway system in the San Fernando Valley, California, during the earthquake of 9 February 1971 (*U.S. Geological Survey photograph*). (*b*) Lower Van Norman dam, California, during same earthquake (*U.S. Geological Survey photograph*). (*c*) Statue of Louis Agassiz in the patio at Stanford University following the 1906 San Francisco earthquake—the search for knowledge goes on (*Courtesy Carnegie Institution of Washington*).

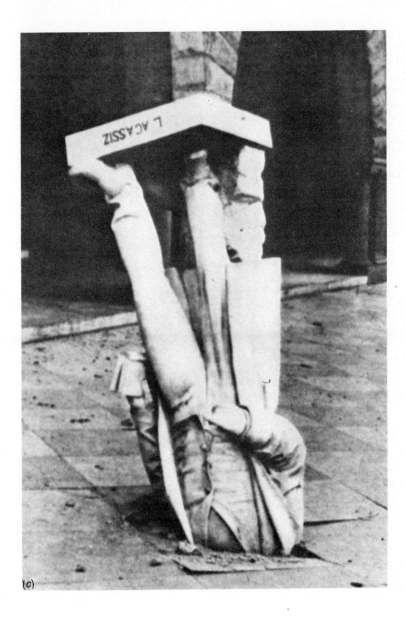

(c)

18

PLANET EARTH

Table 2-2   Modified Mercalli Scale of Earthquake Intensities with Approximately Corresponding Magnitudes. (After Holmes)

Intensity	Description of characteristic effects	Magnitude corresponding to highest intensity reached
I	Instrumental: detected only by seismographs	
II	Feeble: noticed only by sensitive people	3.5–4.2
III	Slight: like the vibrations due to a passing lorry; felt by people at rest, especially on upper floors	
IV	Moderate: felt by people while walking; rocking of loose objects, including standing vehicles	4.3–4.8
V	Rather strong: felt generally; most sleepers are wakened and bells ring	
VI	Strong: trees sway and all suspended objects swing; damage by overturning and falling of loose objects	4.9–5.4
VII	Very strong: general alarm; walls crack; plaster falls	5.5–6.1
VIII	Destructive: car drivers seriously disturbed; masonry fissured; chimneys fall; poorly constructed buildings damaged	6.2–6.9
IX	Ruinous: some houses collapse where ground begins to crack, and pipes break open	
X	Disastrous: ground cracks badly; many buildings destroyed and railway lines bent; landslides on steep slopes	7–7.3
XI	Very disastrous: few buildings remain standing; bridges destroyed; all services (railways, pipes, and cables) out of action; great landslides and floods	7.4–8.1
XII	Catastrophic: total destruction; objects thrown into air; ground rises and falls in waves	>8.1 (maximum known, 8.7)

sequence. The intensity is measured relative to the effects of the earthquake on humans, buildings, pipes, bridges, etc. Also in this table are listed the corresponding *magnitudes* of earthquakes. The magnitude scale for an earthquake, originally defined by Richter in 1935, is essentially a measure of the energy released by an earthquake. It is important to remember that the magnitude scale follows a *logarithmic law*. Thus, in going from any given magnitude to the next one higher, the associated earthquake energy release is multiplied by about 30.

Some feeling for the actual energies involved may be derived by considering that the biggest modern earthquake, which occurred in 1952 in Assam, India, had a magnitude of 8.7, corresponding to an energy release of about 6,000 times that unleashed by the atomic bomb which devastated Hiroshima. The earthquake which ruined San Francisco in 1906 was of magnitude 8.3, while the great Alaskan earthquake of 1964 was of magnitude 8.6. A curious feature of the energy released by earthquake activity is that as much as 80 percent of all the energy liberated in a year comes from the one or two giant earthquakes whose magnitude exceeds 8.0.

Distribution of Earthquakes in Space and Time

The distribution of earthquakes over the earth is far from random. It may be seen at a glance from Fig. 2-2 that the vast majority of earthquakes occur along three long belts—the circum-Pacific belt, in which about 80 percent of the earth's earthquake energy release takes place; the globe-encircling mid-oceanic ridge system; and the *continental fracture system*, which runs from the East Indies, through the Himalayas, southern Asia, and the Mediterranean, finally perhaps, running out to sea to meet the mid-oceanic system near the Azores. Now, earthquakes occur at all levels within the earth down to a depth of about 750 km. None has been recorded at a greater depth than this. When the epicentres of those earthquakes with foci in the 100 to 700 km range are plotted, as in Fig. 2-3, we see that these *intermediate* and *deep* earthquakes are heavily concentrated in certain portions of the circum-Pacific belt. Furthermore, interesting spatial relationships exist among this group of disturbances. Thus many of them are associated with large arcuate structures, and in fact the earthquake foci are found to lie along planes dipping under the arcs, as shown for the Kurile-Kamchatka arc in Fig. 2-4. In contrast, the mid-oceanic shocks are all shallow and, it is clear, are evidently produced by a different mechanism. These features

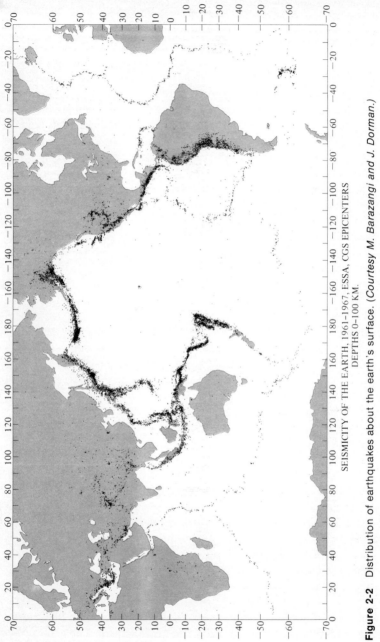

SEISMICITY OF THE EARTH, 1961–1967, ESSA, CGS EPICENTERS
DEPTHS 0–100 KM.

Figure 2-2 Distribution of earthquakes about the earth's surface. (*Courtesy M. Barazangi and J. Dorman.*)

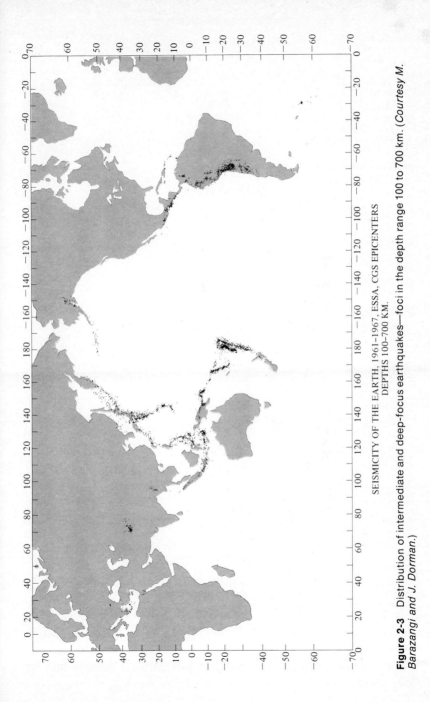

SEISMICITY OF THE EARTH, 1961-1967, ESSA, CGS EPICENTERS
DEPTHS 100-700 KM.

Figure 2-3 Distribution of intermediate and deep-focus earthquakes—foci in the depth range 100 to 700 km. (*Courtesy M. Barazangi and J. Dorman.*)

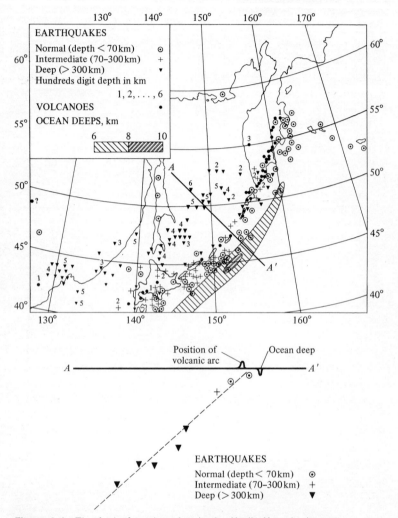

Figure 2-4 The foci of earthquakes in the Kurile-Kamchatka arc have been observed to lie along a plane dipping into the earth at an angle of 40° to the horizontal. (*After B. Gutenberg, C. F. Richter, and J. H. Hodgson.*)

of distribution and type of earthquake are fundamental to the theory of *plate tectonics*, and we will look at them again in the final chapter.

The frequency of occurrence of earthquakes is well illustrated by Table 2-3. Evidently there is one shock per minute of the order of magnitude 2 or greater. While the vast number of earthquakes are small ones, we should remember our earlier comment that the great preponderance of the energy released by earthquakes per year is concentrated in the few large shocks.

Focal Mechanisms

An earthquake is a burst of energy produced by the sudden release of stress within the earth. The state of stress is built up by the forces which mould the features of the earth's surface. Such forces and the major motions they produce over millions of years at the earth's surface will be discussed in Chap. 6 and so will not be described here. We can, however, say a little about the probable mechanism of stress release. Some (not all) earthquakes are clearly associated with movement along visible *faults* in the ground. Faults are merely breaks or discontinuities in the earth's crust. They may have depths of tens of kilometres and may extend over hundreds of kilometres. Following earthquakes associated with such faults, displacements of several metres between points on opposite sides of the fault are sometimes seen. The relative displacements may be purely vertical or horizontal or any combination of these. A beautiful illustration of the result of a horizontal slip along the San Andreas fault in California is shown in Fig. 2-5. A fault showing the result of vertical motion is given in Fig. 2-6. An analysis of measurements of the motion which took place along the San Andreas fault during the 1906

Table 2-3   Frequency of Earthquake Occurrence.
(After Gutenberg and Richter, 1954)

	Magnitude M	Number per year
Great earthquakes	8	1.1
Major earthquakes	7–7.9	18
Destructive earthquakes	6–6.9	120
Damaging earthquakes	5–5.9	800
Minor earthquakes	4–4.9	6,200
Smallest generally felt	3–3.9	49,000
Sometimes felt	2–2.9	300,000

Figure 2-5 A clear record of an earthquake-produced horizontal slip is preserved by the displaced rows of fruit trees in this orchard growing across the San Andreas fault in California. (*Courtesy David E. Scherman, Time-LIFE Picture Agency.*)

Figure 2-6 There may also be considerable vertical motion along a fault during an earthquake. This feature at Hanning Bay was produced by up to 15 ft of vertical displacement during the great 1964 Alaska earthquake. (*Courtesy G. Plafker, U.S. Geological Survey.*)

earthquake that wrecked San Francisco resulted in H. F. Reid's (1911) proposing the elastic-rebound theory of earthquakes, which is still considered valid. Reid's theory may be summarized as follows:

1 Relative displacement of adjacent portions of the earth's crust produces strains greater than the rock can stand. The consequent fracture of the rock causes an earthquake.

2 These relative displacements build up gradually over a long time.

3 The sudden elastic rebounds of the sides of the fracture towards their positions of no strain are the only mass movements at the time of the rupture.

4 The earthquake vibrations spring from the surface of fracture. This surface area is initially small but may quickly expand.

5 The energy released by the earthquake is derived from the strain energy stored in the rock before the shock.

The process is shown schematically in Fig. 2-7. Many earthquakes are not accompanied by a break in the earth's surface, but this model, or something approaching it, is envisaged to apply to most earthquakes. The rupture at the focus has been calculated to spread with velocities of about 3.5 km/second in some cases.

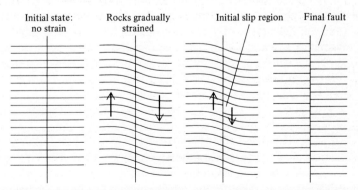

Figure 2-7 Stages in the buildup and occurrence of an earthquake as postulated by Reid. (*After H. Benioff.*)

SEISMIC WAVES AND THE EARTH'S INTERNAL STRUCTURE

The French mathematician Poisson in 1829 showed that two different kinds of waves can travel through the *interior* of an elastic body. These are (1) waves which in their passage through the material cause it to be alternately compressed and expanded—exactly like sound waves in the air—and (2) distortional waves in which the shape of every miniscule element of the body is distorted but the volume of the element remains unchanged. At sufficiently large distances from the earthquake focus (and we shall assume this situation hereafter) the two wave types can be considered as *longitudinal* and *shear* waves. With the former, the particles of the earth may be regarded as vibrating to and fro along the direction in which the wave is travelling. In contrast, with shear waves the particle vibrations are always *at right angles* to the direction of propagation. The distinction may be easily seen in Fig. 2-8. In seismology the longitudinal waves are called P waves and the shear waves are designated S. Perhaps the chief feature of these waves is that the P waves travel faster than the S type. Therefore, following an earthquake, the P vibrations will always reach a seismic recording station first (the letter P, in fact, stands for primary) followed by the S vibrations (the letter S standing for secondary). This effect is shown for an actual earthquake in Fig. 2-9.

P and S waves are known as *body* waves, travelling as they do through the body of the earth. In 1885 Lord Rayleigh proved that another type of wave, a *surface* wave, is possible. The *Rayleigh wave* is similar to the familiar waves seen on water surfaces. The elliptic motion of a surface particle during the passage of a Rayleigh wave is shown in Fig. 2-10a. The last really different earthquake-generated wave was found in 1911 by Love, who showed that a transverse wave could hug the earth's surface, provided there was a level within the earth at which the elastic properties changed. This is illustrated in Fig. 2-10b. Both types of surface wave travel more slowly than the S waves and so are received at seismic stations later than P and S arrivals. Lord Rayleigh made the point that the body waves (P and S) spread out

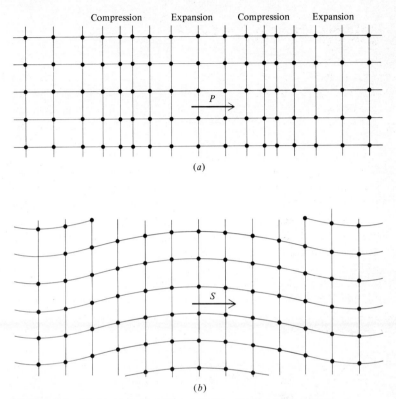

Compression   Expansion   Compression   Expansion

(*a*)

(*b*)

Figure 2-8 (*a*) In *P* waves the particles of the earth vibrate backwards and forwards along the direction in which the wave advances. (*b*) The direction of particle vibration during the passage of *S* waves is at right angles to the direction of the wave's travel. (*After G. S. Kino and J. Shaw.*)

through the earth in three dimensions, whereas surface waves spread only in two dimensions and therefore lose energy more slowly. The surface-wave signals, as a result, are in fact often found to be considerably larger than those of *P* and *S* (see Fig. 2-9).

In the 1890s Milne developed a suitable seismograph which could be installed in many parts of the world, and in 1900 Oldham first identified the *P*-, *S*-, and surface-wave arrivals on a seismogram. Since then an increasingly detailed and systematic study

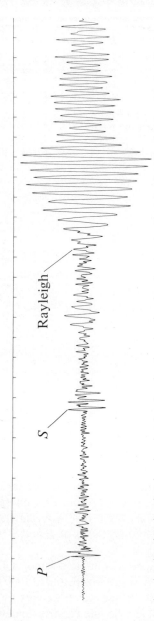

Figure 2-9 The arrival of seismic waves at an observatory is recorded by the vibration of a pen writing on a slowly moving chart paper. The first arrival is the *P* wave, followed by the *S* wave. The more slowly moving surface waves appear later, but they are frequently the biggest signal received. Recording taken at Revelstoke (Canada) following a magnitude 6.1 earthquake in the Kuriles. *(Courtesy A. J. Wickens, Department of Energy, Mines and Resources, Government of Canada.)*

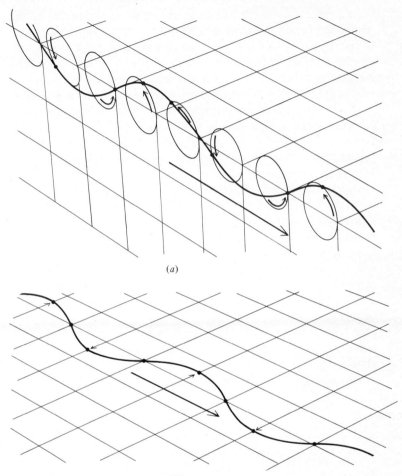

(a)

(b)

Figure 2-10 (a) During the passage of a Rayleigh surface wave, particles describe ellipses. (b) Particle motion in the Love surface wave is perpendicular to the direction of wave propagation. (*After J. Oliver.*)

of the times of arrivals of these waves (after an earthquake) at as many points of the earth's surface as possible has led to a remarkably finely drawn picture of our planet's interior.

Seismic waves are like light waves in many ways, and the paths they follow are governed by laws similar to those of optics. In particular, to put it simply, we can say that seismic and light vibrations both try to get from any point A to another point B by the quickest route. If, then, the earth were a perfectly homogeneous body in which seismic waves travelled at a constant speed, seismic rays would be straight lines. Obviously under these circumstances a straight-line path would be the quickest way between two points. We show such straight travel paths for seismic rays in an ideally simple uniform earth in Fig. 2-11. However, in the early seismic investigations it was found that vibrations from earthquakes were arriving earlier and earlier than they should on this simple model at observatories which were successively farther and farther from the earthquake source. But, as may be seen from Fig. 2-11, rays reaching distant observatories have penetrated deeper into the earth than those reaching near observatories. Obviously then, if these long-distance rays are arriving steadily earlier, it must be that they can travel more quickly as they go down into the earth. In this way, then, it was discovered that the earth is not homogeneous and that its elastic properties change with depth, and so the velocities of *P* and *S* waves increase steadily (with some reservation to be made shortly) at least for many hundreds of kilometres down. As a consequence, seismic rays are curved upwards, as in Fig. 2-12.

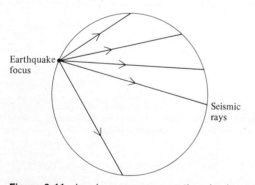

Figure 2-11 In a homogeneous earth, seismic rays would travel in straight lines.

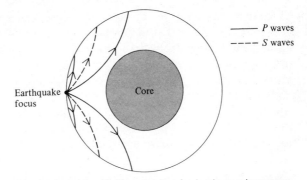

Figure 2-12 In the real earth the velocity of seismic rays increases, usually, with depth, and this effect causes seismic rays to curve upwards as shown.

In 1906 Oldham made another major discovery. He found that following an earthquake, P waves arrived at an observatory at the diametrically opposite point of the earth from the rupture, significantly later than he expected from the times of arrival of the P waves at observatories much closer to the earthquake focus. The most reasonable way to reconcile these time differences is to postulate that the earth has a central core within which the P waves travel relatively slowly. Waves travelling directly to the antipodes of the earthquake would naturally have to travel through this *slow core* region and would be delayed. This suggestion of a core by Oldham harmonized well with speculation in the ninetenth century that the earth might have a dense core. This latter speculation was based on the fact that whereas rocks found at the earth's surface are about 3 times as dense as water, the average density of the whole earth had been shown by Cavendish to be 5.5 times that of water. Evidently there had to be some region of high density within the earth, and this could well correspond to Oldham's core.

Oldham made the further interesting suggestion that a *shadow zone* of P-wave arrivals would be found because of the abrupt fall in P-wave velocity at the surface of the core. To explain this, we turn to Fig. 2-13a. Here we see how seismic energy spreads through the earth following an earthquake. As we look at rays which leave the focus at steeper and steeper angles to the surface of the earth, we see that they reach the earth's surface at points farther and farther from the earthquake. Everything progresses smoothly until we reach rays which reach the earth's surface about 105° distant from the earthquake focus. We then come to ray 1 which just touches the core. Vibrations are, therefore, sent into the core, as shown in Fig. 2-13a. The ray follows the curve shown in the core and is then bent sharply on emergence from the core so that ray 1 actually emerges at about 185° away from the source, i.e., over halfway around the earth from the source. The uniform progression of seismic arrivals around the earth is thus broken at 105°. We next concentrate on Fig. 2-13b from which we have omitted all the earlier rays, except ray 1, for clarity. The next rays, going more steeply into the earth than ray 1, are not so severely deflected in total by the core as is ray 1 so that they emerge at the surface at angular distances of less than 180° from the focus. Ray 2 illustrates one of these rays. This effect proceeds until ray 3, which emerges at about 143°, is reached. Steeper rays than this are bent more severely by the core, as shown in Fig. 2-13c. Hence we have a shadow zone between 105° and 143° where the direct P waves are absent. Our diagrams, of course, have all been of sections through the earth. When we consider that the earth is a three-dimensional sphere rather than a circle, we see that the shadow zone is a band on the earth's surface, as we have shown in Fig. 2-13d. Oldham's prediction of a shadow zone was confirmed from examination of seismograms by Gutenberg in 1914. Gutenberg found the shadow zone location relative to the earthquake focus and then calculated that the depth to the core surface must be 2,900 km. This has proved to be a remarkably precise estimate. A quarter of a century later (1939), Jeffreys analysed a much greater and more reliable number of data and concluded the core was at a depth of 2,898 km, with an uncertainty of 4 km. In the past 5 years or so, much work has been done on the observations of the great pulsing vibrations of the whole earth triggered by massive earthquakes. These suggest that

Figure 2-13 (a) to (d) The earth's core deflects seismic rays in such a way as to produce a shadow zone on the earth's surface. The sequence of diagrams is explained in the text. [(a) to (c) after B. Gutenberg; (d) after A. Holmes.]

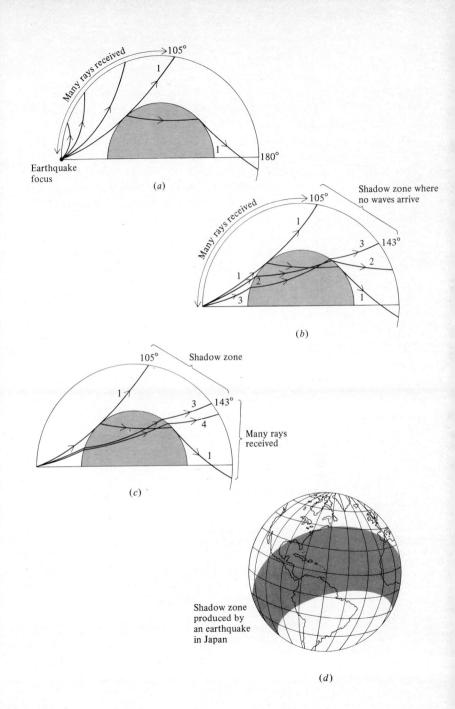

(a)

Many rays received
105°
1
180°
Earthquake
focus

(b)

Many rays received
105°
1
Shadow zone where
no waves arrive
143°
3
2
1
1
2
3

(c)

105°
Shadow zone
1
143°
3
4
Many rays
received
1

(d)

Shadow zone
produced by
an earthquake
in Japan

perhaps Gutenberg's value *may* be up to 20 km too large, a possible error of about 2/3 of 1 percent.

The shadow zone is not totally devoid of P arrivals. Low-amplitude P waves are received throughout this region and for a number of years were not completely understood. It was appreciated that some were due to the effect known as *diffraction* in the theory of waves. This merely means that waves do not cast perfect shadows and will always curve into the shaded region. However, in 1935 I. Lehmann pointed out that some of these shadow-zone arrivals would be better explained by allowing the earth's core itself to have a central core in which the P waves travel more quickly than in the outer core. This central core could then deflect deeply penetrating P waves out into the shadow zone, as we show in Fig. 2-14. In 1939, Jeffreys proved theoretically that all the small P waves arriving in the shadow zone could not be due to diffraction, and Lehmann's hypothetical division of the core into two regions became widely accepted, the outer boundary of the inner core occurring at a depth of about 5,120 km.

So far we have seen how a picture was developed of the earth as a sphere with a core 2,900 km down, which itself has a

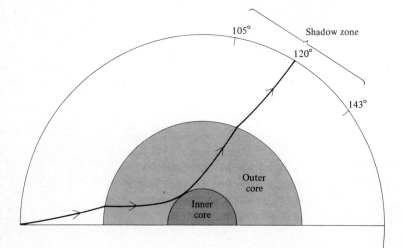

Figure 2-14 The deflection of P waves by the inner core into the shadow zone led to the discovery of the inner core by I. Lehmann.

small inner core. The earth outside the core came to be known as the *mantle*. Meanwhile, considerable progress was being made in the determination of the details of the structure of the outer 50 km of the earth. In 1909 A. Mohorovicic found the first seismic evidence that the earth has a crust which is different from the mantle. Seismic records taken within a few hundred miles of a Yugoslavian earthquake showed clearly that two *P* and two *S* waves were radiating away from the source. Mohorovicic correctly interpreted this to be due to the existence of a surface layer on the earth in which the waves travelled more slowly than in the mantle. Waves travelling in the top layer from the source to the receiver cover a shorter path than those which go down through the layer to the mantle, move quickly through the mantle, and then emerge through the top layer to the receiver. Thus at short distances from the source the receiver records the waves through the top layer before those which spent part of their journey in the mantle. Before too long, however, despite their longer physical path, the deeper travelling waves make up so much time in travelling at the faster mantle velocity that at further receiving stations they are now the first phase to arrive (see Fig. 2-15). At greater and greater distances the mantle wave gets farther and farther ahead. Mohorovicic's top layer is now recognized as the crust of the earth, and the boundary between the crust and the mantle is called the *Mohorovicic discontinuity* or, frequently, just the *Moho*. This part of the crust was calculated by Mohorovicic to be about 54 km thick, a value which was later found to be too great.

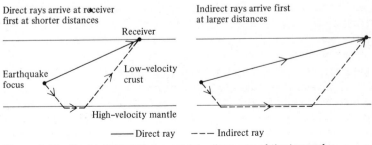

Figure 2-15 Illustration of Mohorovicic's discovery of the base of the crust.

The Mohorovicic discontinuity has subsequently been shown by seismologists to be a global phenomenon absent in only a few areas. The crust, that is, all rock above this discontinuity, falls into two distinct categories—continental and oceanic—and we will look briefly at these in turn.

The continental crust, on average, is about 40 km thick. In many areas a coating of sedimentary and volcanic rocks, sometimes thousands of feet thick, is found. Underlying this, or exposed at the surface in the shield areas, we find granitic and metamorphic rock. The detailed structure of the continental crust has, however, not been well established. A subdivision of the crust was found in 1925 by Conrad, who, in examining earthquake records taken fairly close to a shock in Austria, found a *P* wave in addition to the two *P*'s and two *S*'s observed by Mohorovicic. He therefore suggested that Mohorovicic's crust was really composed of two layers. This *Conrad discontinuity* has, however, proved to be more elusive than the Mohorovicic discontinuity and has not been found in all continental areas. Recent seismic studies indicate that the velocities of seismic waves and rock densities increase with depth into the crust in such a way that while the outer portion is granitic in type, the lower crust is of *intermediate composition*, being intermediate between the granitic and the basaltic types found under the oceans. We know from heat-flow considerations that the whole crust cannot have the composition of granite. This is because the relatively high concentrations of the radioactive elements U, Th, and K in granites would produce a far higher flow of heat from the continental crust than is actually observed.

The oceanic crust contrasts strongly with the continental. It is much thinner—5 to 12 km thick—and has no granitic layer. Its structure also appears simpler. Three layers are usually identified above the Mohorovicic discontinuity. Layer 1 is a thin sedimentary veneer usually less than 1 km thick. Layer 2 is about 4 km thick and is considered to be basaltic. The identity of layer 3, which is about 1 km thick, is not clear, but it is probably metamorphosed basalt and gabbro.

Figure 2-16 shows the interior structure of the earth as we have just developed it, while in Fig. 2-17 a larger scale representa-

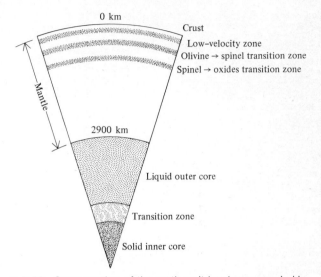

Figure 2-16   Cross section of the earth as it has been revealed by seismology.

tion is made of typical continental and oceanic crustal cross sections. The divisions shown in Fig. 2-16 are all boundaries at which seismic velocities change abruptly. To see the magnitude of the velocities involved and the changes, we turn to Fig. 2-18 where are drawn the variations with depth of the *P*- and *S*-wave velocities in the earth. The most dramatic variation, of course, is seen at the core boundary at a depth of about 2,900 km. It is all the more striking because the *S*-wave velocity curve stops completely here. No earthquake record has ever been reliably interpreted as involving an *S* wave travelling through the outer core. Since solids transmit shear (*S*) waves but liquids do not, it is, therefore, considered that the outer core is liquid. The inner core is probably solid, being essentially just the same material as the outer core but having solidified under the enormous pressures near the very centre of the earth. Another feature which has attracted much attention since Gutenberg first commented on it in 1926 is the *low-velocity zone* in the upper mantle, lying at a depth of around 150 km and being perhaps 100 km thick. In this layer the normal trend in the mantle of velocities to *increase* with depth

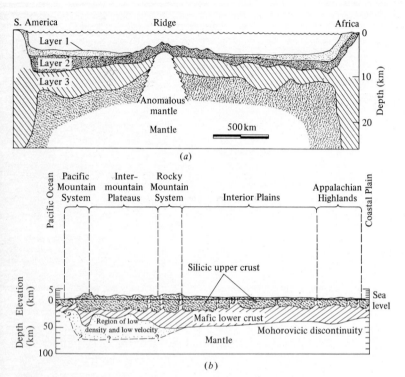

Figure 2-17 (a) Typical oceanic crust cross section. (b) Continental crust cross section of U.S.A. (*After L. C. Pakiser, I. Zietz, and P. J. Wyllie.*)

is reversed. It appears as if at this level the tendency of increasing pressure to increase seismic velocities is overcome by the reduction of velocities caused by increasing temperature. We will see in the final chapter that this low-velocity zone plays a key role in plate-tectonic theories of motions at the earth's surface.

So far, we have been dividing up the earth on the basis of seismic velocity variations. Adams and Williamson pointed out in 1925 that the density at points within the earth could be reasonably well calculated from the P and S velocities at the same points. Such computations were done in great detail with added refinements over many years, principally by Bullen. We see from Fig. 2-19 that the density ranges from approximately 3 in the crust

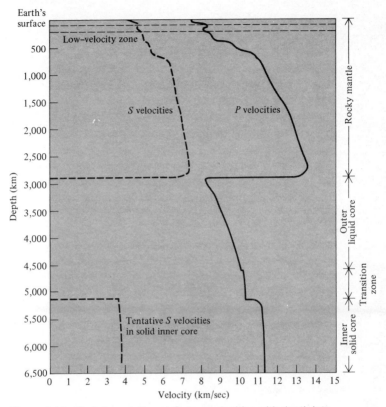

Figure 2-18 Variation of *P*- and *S*-wave velocities with depth into the earth's mantle and core. (*After B. A. Bolt.*)

to about 6 at the base of the mantle. Within the core, the density jumps to about 9.5, steadily rises to about 12 at the inner core, where it jumps slightly to about 13.5, and remains constant in the inner core. (We are using density units in which the density of water is 1.) Comparison of these density and elastic-property variations in the mantle with laboratory measurements on rocks of various compositions by Jeffreys, Bullen, Birch, Anderson, and others led to the conclusions that the upper mantle material is chemically different from the crust, being denser and richer in the oxides of iron and magnesium relative to silicon oxide. The most

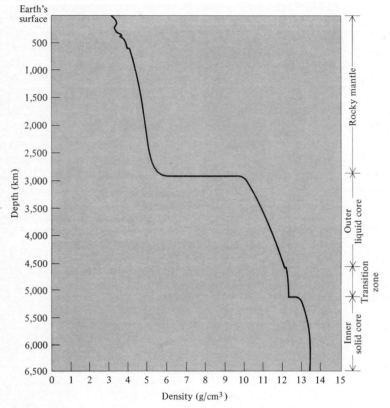

Figure 2-19   The variation of density within the earth. (*After B. A. Bolt.*)

important mineral in the upper mantle is probably olivine, a magnesium-iron silicate. In the upper mantle we find the previously mentioned low-velocity zone. At greater depth the seismic velocities and density increase reasonably, as one might expect from the effects of increasing pressure. At about 350 km depth, however, we see, in Fig. 2-18, a clear increase in the rate of increase of velocities, and another such increase occurs at about 700 km depth. These discontinuities have been defined in recent years by Johnson, Anderson, and others and have associated with them sharp density increases (Fig. 2-19). The change at 350 km is thought to be due to a *phase change* in the mineral olivine. In this transition the atoms comprised in olivine rearrange themselves into a more compact "spinel" structure in response to the high pressure at this depth. That this olivine-spinel transition might occur in the mantle was first suggested by Bernal in 1936. The discontinuity at 700 km probably represents another pressure and temperature zone in which mineralogical changes occur. It is thought that here, mantle minerals break down into very compact oxide structures. These forms are probably maintained down to the core. The outer core is generally considered to be liquid iron, with some nickel and perhaps small amounts of other elements like silicon and sulphur. What is known of the physical-chemical properties of iron at high pressures and temperatures is consistent with this. The inner core, as mentioned earlier, is probably just a solidified version of the outer core. While this general picture is now fairly widely accepted, it is worth recalling that scientists have at times suggested that the core might be made of hydrogen under extreme pressure or that it might be of the same chemical composition as the lower mantle but in the more dense phase. Modern high-pressure studies do not seem to support such models, however.

In the past decade seismology has been greatly aided by the electronic revolution, and techniques have improved enormously. Magnetic-tape recording, digital techniques, and arrays of seismographs have been employed. Seismologists in many countries have been measuring with ever more precision the depth to the Mohorovicic discontinuity and to the outer and inner cores. The details of the variation in seismic velocity in the mantle are being studied, the low-velocity layer is being mapped, and lateral variations in the mantle have been found. Most of the details of Fig. 2-18 were originally developed in studies of the P and S body waves. Now upper mantle features are being scrutinized in surface-wave studies, and earth models are tested against the observed vibrations of the whole earth, which are recorded following major earthquakes. Details of the core structure are being examined with the aid of rays such as the $P5KP$, shown in Fig. 2-20, which travels as a P wave through the mantle, is reflected 4 times at the core-mantle boundary, and then traverses

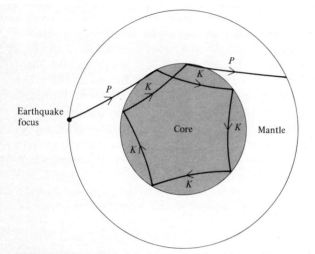

Figure 2-20 A *P5KP* wave is a *P* wave throughout its travels, and within the liquid outer core is reflected 4 times. Studies of such multiply reflected waves are important in efforts to establish whether there is some topography on the core-mantle boundary.

the mantle to emerge at the earth's surface, having been a *P* wave throughout the trip. While it seems highly unlikely any serious changes will be made to the velocity distributions of Fig. 2-18, undoubtedly much interesting fine detail remains to be explored.

Chapter 3

Gravity and the Figure of the Earth

Ever since Newton's discovery of the inverse square law of gravity, the earth's gravitational field and its vagaries have been the subject of considerable interest. Much may be, and has been, learnt from a study of the wrinkles in the gravity field over the earth's surface. The question of the shape of the earth is inevitably involved in such investigations, and we consider this question first, in the next section. The information provided by the earth's gravitational field on the planet's internal structure and its approach to hydrostatic equilibrium is discussed later in the chapter.

THE EARTH'S SHAPE

Eratosthenes' Sphere

Seen from the moon, the earth appears to be a circular disc, and in fact, distant observers would consider the earth to be reason-

Figure 3-1 View of the earth from the moon showing the close approach to circularity of the profile of our planet. (*NASA photo.*)

ably spherical. Impressive mountain ranges such as the Himalayas would be only the merest distortions (Fig. 3-1).

Pythagoras and his school were the first to appreciate the rotundity of the earth, in the sixth century B.C., and Eratosthenes even carried out a fairly accurate calculation of the earth's radius about 300 years later. His method, while not original, was elegant and is worthy of recall. Eratosthenes was aware that at noon, at mid-summer, the sun was immediately overhead at Syene (now Aswan) since it shone directly into a deep well. It was known, however, that at the same time the sun was 7.2° out of the vertical at Alexandria, which is virtually due north of Syene. Hence, from the elementary geometry of Fig. 3-2 we see that the radius of the earth is given by dividing the distance from Syene to Alexandria by 7.2° when the latter is expressed in radians (0.125 radian). Now Eratosthenes estimated the separation of Syene and Alexandria in the units of that era as about 5,000 stadia, basing his estimate on the time taken by camel trains to cover the distance, whence he calculated that the radius of the earth equaled 5,000/0.125 stadia. Because of some uncertainty as to the definition of the stadium unit this could correspond to either 7,330 km or 6,340 km. Either value is remarkably close to the presently accepted average radius of 6,371 km, and the computation must be regarded as a remarkable intellectual feat. The method has been used many times since, with refinements, and the earth's radius may now be accurately determined in this way.

Newton's Spheroid

In the *Principia*, published in 1687, Newton pointed out that, owing to the earth's daily rotation about its N-S axis, centrifugal forces would be expected to produce an equatorial bulge of the earth. The earth would thus look like a flattened sphere, or a *spheroid* (see Fig. 3-3). In his remarkable way, Newton went on to calculate the flattening of the earth. This is defined as the ratio $f = (a-b)/a$, where a equals the equatorial radius of the earth and b equals the polar radius. Newton obtained the value of 1/230 for the flattening. That is to say, he thought the equatorial radius exceeded the polar radius by 1 part in 230.

Numerous attempts were subsequently made to verify ex-

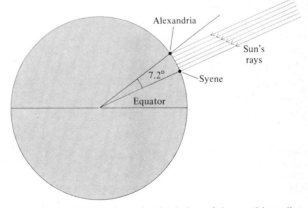

Figure 3-2 Erathosthenes' calculation of the earth's radius.

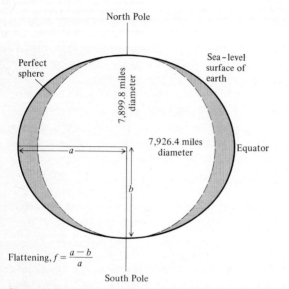

Figure 3-3 Newton pointed out that centrifugal forces caused by the earth's rotation would create a flattening at the poles and a spare tire around its middle. In fact the equatorial radius exceeds the polar radius by a little less than 1/3 of 1 percent. (*After D. G. King-Hele.*)

perimentally this predicted flattening. If the earth were a perfect sphere, a degree of latitude would correspond to a constant length of an arc on the earth's surface. Should the earth, however, be flattened at the poles, then we would expect the arc length at the pole corresponding to 1° of latitude to be greater than the equivalent arc length at the equator (Fig. 3-4). A number of expeditions set off in the eighteenth century to test such speculations. These expeditions were often long and extraordinarily arduous. Not surprisingly there was some early confusion. The

Figure 3-4 On a perfectly spherical earth, 1° of latitude would correspond to a fixed arc length on the surface wherever it was measured. If the earth were flattened at the poles, however, 1° of latitude measured near the poles would correspond to a greater surface arc length than that corresponding to 1° measured at the equator.

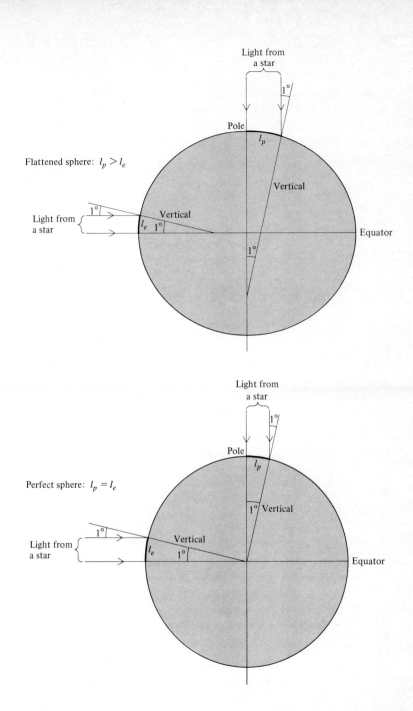

Light from
a star

1°

Pole

l_p

Flattened sphere: $l_p > l_e$

Vertical

Light from
a star

1°

Vertical

l_e 1°

Equator

1°

Light from
a star

1°

Pole

l_p

Perfect sphere: $l_p = l_e$

1° Vertical

Light from
a star

1°

Vertical

l_e 1°

Equator

French astronomer Cassini and his son actually concluded that
the earth was flattened at the equator and bulged at the poles,
having obtained a flattening of $-1/95$. Subsequent estimates are
given in Table 3-1 taken from King-Hele (1966). It is evident that
Newton's proposal of an equatorial bulge has been amply verified. The flattening, however, is not as large as he estimated. The
methods used to estimate f included not only comparisons of
arc-length measurements but also studies of perturbations of the
moon's motions, examination of the precession of the earth's
rotation axis, and detailed studies of the variation of gravity over
the earth's surface.

The launching of *artificial* satellites since 1957 has now
provided a new and much more powerful tool with which to
investigate the earth's shape. In the space of a few years, studies
of satellite orbits have allowed the flattening to be calculated far
more accurately than ever before and have also provided an
enormous amount of new information about distortions smaller
than the equatorial bulge in the earth's spherical shape. It is no
exaggeration to say the artificial satellites have revolutionized our
knowledge of the earth's figure.

The Sea-Level Surface, or Geoid

So far we have been talking loosely about the earth's shape. To be
more precise, what we mean in fact is the shape of the sea-level
surface all over the world, wave and tide influence being ignored.
For the 70 percent of the earth's surface which is covered by the
oceans this is an intuitively obvious surface to choose. To extend
the concept to continental areas, we have to imagine the land
masses having been dissected by a latticework of small canals
open to the sea. The levels to which the water rose in these
channels would then be the sea-level surface on the continents. It
is this sea-level surface, then, whose flattening was estimated to
be 1/297.1 by Jeffreys in 1948 (Table 3-1). The solid-rock surface,
of course, follows the general trend of the larger features of the
sea-level surface, but is subject to local distortions relative to it

Table 3-1 Estimates of *f*, the Earth's Flattening. (After King-Hele)

Investigator	Date	I/*f*
Newton	1687	230
Huygens	1690	578
Cassini	1720	−95
Maupertius	1738	178
Peruvian party	1748	179–266
Boscovich	1760	248
Legendre	1789	318
Everest	1830	301
Bessel	1841	299
Clarke	1866	295
Hayford	1909	297.0
Heiskanen	1929	298.2
Krassovsky	1938	298.3
Jeffreys	1948	297.1

on account of the existence of such things as mountains. The sea-level surface is often referred to as the *geoid*.

To get some feeling for the magnitude of the earth's equatorial bulge, we might observe that the highest mountains on earth are less than 10 km high, while the difference between the equatorial and polar radii of earth is about 21.5 km (13.4 miles).

SATELLITE INVESTIGATIONS OF THE EARTH'S SHAPE

The Equatorial Bulge

The earth's first artificial satellite, *Sputnik I*, was launched on October 4, 1957. By 1958, Merson and King-Hele had analysed the orbit of *Sputnik II* and redetermined the value of the earth's flattening. They found a value $1/f = 298.24 \pm 0.02$ compared with the widely accepted Jeffreys' value of 297.1 ± 0.4. The earth was thus found to be a little less flattened than had previously been thought, and the flattening was calculated with much higher accuracy.

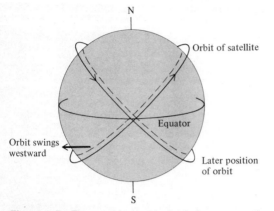

Figure 3-5 The earth's equatorial bulge causes the plane of the artificial satellite's orbit to rotate steadily (precess) about the earth's axis of spin.

If the earth were perfectly spherical and had no lateral variations in density, any earth satellite would have an elliptic orbit of constant dimensions and would remain in a plane fixed relative to the distant stars (neglecting air drag close to the earth and minor perturbations caused by the sun and the moon). The earth, of course, is rotating west to east on its axis, and so to an earthly observer the satellite orbit (just like the fixed stars) appears to be rotating east to west about the earth. Departures of the earth from perfect spherical shape disturb this simple state of affairs. By far the most important departure from sphericity is the equatorial bulge, and this has the greatest disturbing effect on the satellite orbit. The bulge causes two significant things to happen. Firstly, the plane of the satellite orbit no longer remains fixed in space but now rotates steadily about the earth's axis, as much as 5 to 10° per day (see Fig. 3-5). Secondly, the perigee point in the orbit moves in the orbital plane around the orbit with a period which may be anything from a few months to several years (Fig. 3-6). The earth's flattening may be calculated from measurements of both these effects. The most accurate results come from

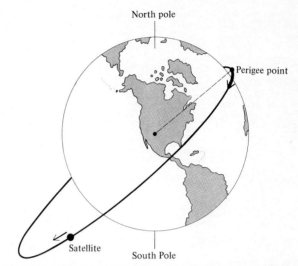

Figure 3-6 The earth's flattening also causes the perigee point of the satellite's orbit to move around the orbit. (*After D. G. King-Hele.*)

studying the rotation of the satellite's orbital plane. Errors in the determination of the angle of rotation may be reduced considerably by allowing the rotation to accumulate over several months, hence many orbits, so that the orbit has rotated several hundreds of degrees. The flattening may then be deduced with an error of about 1 in 10,000.

Measurements of the rate of rotation of the perigee point around the orbit yield less precise results for f, but do not conflict with that based on the rotation of the orbital plane.

The currently accepted value for f based on the analysis of many satellite orbits is $1/298.26 \pm 0.02$.

The Pear Shape

A wholly new and unexpected result of satellite orbit analysis was that the sea-level surface is very slightly pear-shaped with the stem towards the North Pole. It appears that the North Pole sea-level surface is about 40 m further from the equator than is

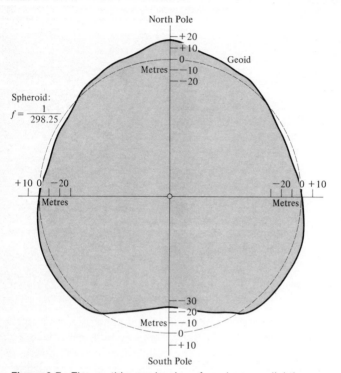

Figure 3-7 The earth's sea-level surface is very slightly pear-shaped. Its distortions are greatly magnified here. (*After D. G. King-Hele.*)

the South Pole surface (Fig. 3-7). The earth's mass is thus not symmetrically distributed about the equatorial plane, an odd result! The "pear-shapedness" causes the satellite's perigee distance to the earth's centre to be less when the perigee is in the Northern Hemisphere than when in the Southern (Fig. 3-8).

Numerous other, smaller distortions of the sea-level surface have been determined from satellite studies. These will not be considered here, although they contribute to the appearance of the contour diagram of the sea-level surface shown in Fig. 3-9.

ISOSTASY

During the first survey of India, in the 1850s, it was noticed that plumb lines in the Himalayas were deflected less from the vertical (by about two-thirds) than would be expected if the mountains were sitting like blocks on the surface of a uniform earth (Pratt, 1855). Since the visible portions of the mountains must certainly have been exerting their calculated deflecting force, then it seemed most reasonable to attribute the low deflections to *deficiencies* of mass beneath the mountains. Airy (1855) sought to explain these observations by a very simple mechanism. He suggested that the earth's crust was a thin layer of constant

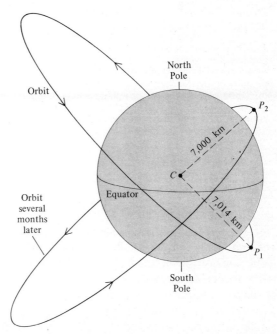

Figure 3-8 The satellite's perigee distance to the earth's centre is less when the perigee is in the Northern Hemisphere (P_2C) than when in the Southern (P_1C) because of the earth's pear shape. (*After D. G. King-Hele.*)

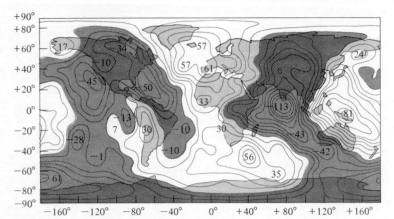

Figure 3-9 The earth's sea-level surface, i.e., the geoid. If the geoid were merely a flattened sphere (a spheroid) with a flattening of $\frac{1}{298.25}$, there would be no need to use any of the contour lines shown. However, the geoid is a more complicated surface in reality, and we show its distortions from an ideal flattened sphere by means of the contour lines which are at 10-metre intervals. Points on the geoid lying on the contour labelled "30" lie at a height 30m above this spheroid, and so on for other positive numbers. Points on the geoid which are actually below the reference spheroid (with $f = \frac{1}{298.25}$) are, of course, labelled with negative numbers. This is the 1969 Smithsonian geoid.

density resting on denser material. Those portions of the crust higher than average he considered were underlain by roots of crustal material dipping deeply into the dense interior. The latter he took to be in hydrostatic equilibrium, and his theory in essence is that the mountains are floating like logs in water. The larger the diameter of the log the higher it protrudes above the surface, but the deeper it reaches. The model is illustrated in Fig. 3-10. Pratt (1859), however, while agreeing with the concept of a light crust floating on a dense interior, disagreed on the details. He considered the base of the crust to be at a fixed depth below sea level. Variations in topographic height were then ascribed to variations in the density in the crust. The situation is shown in Fig. 3-11. Airy considered the total mass under unit area of the crust down to some level beneath the deepest root to be a constant. Pratt's

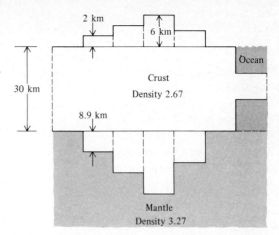

Figure 3-10 Airy's model of how isostatic compensation is maintained. (*After W. A. Heiskanen.*)

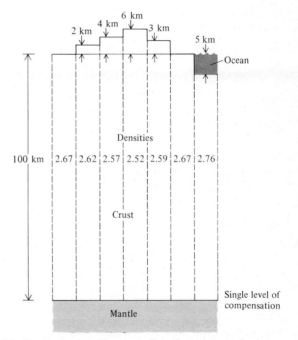

Figure 3-11 Pratt's theory of isostasy. Note that whereas the base of Airy's crust mirrors the topography with magnification, Pratt's crust has a smooth base. (*After W. A. Heiskanen.*)

theory had mass per unit area down to the level of compensation as a constant. Beneath these levels of compensation, hydrostatic equilibrium was supposed to prevail. This set of conditions was said to satisfy the requirements of isostatic equilibrium.

In a century of subsequent testing of these theories of isostasy, it has proved impossible to favour one over the other unequivocally. Variations of both fundamental models fare no better. It is essentially impossible to resolve such questions uniquely by gravitational studies, and it has been necessary to look to seismic studies for assistance. In some instances these have indicated the presence of deep roots beneath high mountains (e.g., the Andes), and generally, shallow crusts are found at sea, as predicted by the Airy model. However, other areas seem more suited to the Pratt hypothesis. Heiskanen (Heiskanen and Vening Meinesz, 1958) suggested that 63 percent of the isostatic equilibrium of the crust is brought about by Airy's mechanism and the remainder by Pratt's. However, regardless of the exact mechanism, all are agreed that topographic highs are generally compensated by low-density underlying material.

The satellite geoid results are a clear indication of isostatic equilibrium on the continental scale. If the continents were sitting like blocks on a uniform earth, they would produce significant upward bulges in the geoid over the continents. Such a correlation between geoidal bulges and continents cannot be seen in Fig. 3-9, and clearly the continents must be compensated by mass deficiencies beneath them.

That the earth as a whole is close to hydrostatic equilibrium is shown by the fact that the measured flattening of 1/298.26 is near to the value one would predict theoretically if the earth were perfectly fluid and were in its equilibrium shape. However, the humps and hollows visible on the geoid in Fig. 3-9 suggest the presence of lateral variations of density within the earth's mantle. If these were *static* features at great depth, they would require considerable shear strength for that part of the mantle in order to resist the tendency of the Archimedes upthrust forces to remove the density variations. Alternatively, they could be dynamic, changing features reflecting the presence of convection currents in the mantle, such as are discussed in Chap. 6. Hot, light, rising limbs of convecting cells would be expected to give hollows on the geoid, while cooler, denser, descending limbs would produce

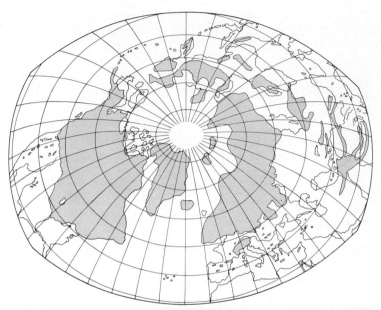

Figure 3-12 The hatching indicates the maximum extent of the glaciation in the Northern Hemisphere. This glaciation ended a mere 7,000 years ago. (*After E. Antevs and R. F. Flint.*)

humps. It is remarkable that two such incompatible states of the deep mantle would both satisfy the geoid picture, as G. D. Garland has emphasized.

Isostasy in Action

It may well be said that we live in an ice age, for the earth enjoyed a mild climate for over 200 million years until about $2^1/_2$ million years ago when a noticeable cooling set in. Since then large areas of the earth have been covered by ice 4 times, and each time the ice has receded. In between these major advances of the ice there have been a number of smaller glaciations. During the major icing episodes as much as one-third of the earth's continental area has been under ice. The last major glaciation ended in North America a mere 7,000 years ago. Before this massive thaw, the site on which is now found the Canadian city of Toronto was under an ice blanket over 1 mile thick. The extent of this glaciation of the Northern Hemisphere is shown in Fig. 3-12. Whether this most

Figure 3-13 Raised beaches above Hudson Bay, Canada. Each of these lines of beaches was originally cut at sea level and was subsequently carried way above sea level as the crust rebounded following melting of the ice load. (*Courtesy Department of Energy, Mines and Resources, Government of Canada.*)

recent cooling will be the last for tens of millions of years or whether the ice will be inching inexorably back in a few tens of thousands of years is completely uncertain at the moment.

Our interest in this phenomenon, however, is not so much in the Ice Age itself as in the remarkable example it provides of isostasy in action, for a continental ice sheet 1,000 miles square and 1 mile thick weighs over 3 million billion tons and puts an enormous load on the continental crust. What happens as a result may be inferred from Airy's model in which the continents resemble logs floating in water. Clearly if we stepped gently onto a floating log it would float a little deeper in the water in order that more water might be displaced so that, by Archimedes' principle, the extra upthrust thus created would support our weight on the log. In the same way we expect that an enormous ice load placed on a continent would press the continent further into the denser mantle to support the ice. Then when the ice melts and this load is removed, we would expect the continent to float back up slowly to its original level, just as the log would resume its original floating depth when we finally lost balance and fell into the lake. That this sequence of events happened during the Ice Age has been verified by geological studies in Scandinavia and North America. Thus at the height of the last glaciation an enormous slab of ice, perhaps 8,000 ft thick, was centred on the Hudson Bay area of Canada and depressed the crust several hundreds of metres into the mantle. By approximately 7,000 years ago this load had completely melted away. The subsequent rise of the North American crust in search of its original equilibrium position is now beautifully recorded by a series of beaches around Hudson Bay. These beaches rise like a series of steps above the current beach which is of course at sea level since Hudson Bay is open to the oceans. A striking photograph of this feature is shown in Fig. 3-13. Carbon-14 dating has shown that these beaches have all been cut by the Hudson Bay waters in the past 7,000 years. Furthermore, the highest raised beach was cut before all the others, and there is a measured steady progression of ages of formation as one goes down these beach steps to the current sea-level beach, which is, of course, the most recently sculpted beach (Fig. 3-14). These results, therefore, are completely consistent with the concept that the continent, having been depressed by the ice load, has been steadily rebounding towards its old position, carrying beaches that were once at sea level to hundreds of feet above the sea. From Fig. 3-14 we can estimate the rate at which the continent has been recovering its balance. We see that the highest beach in this area is now about 150 m above sea level. But evidently the beach was formed at sea level about 7,000 years ago. Therefore we conclude that the land has risen 150 m in 7,000 years, i.e., at a rate of about 2 cm/year, if we assume a constant rate of rebound. While we would, in practice, not expect a constant rebound rate, the 2 cm/year value is a reasonable

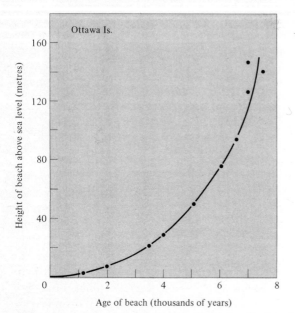

Figure 3-14 The graph shows how the higher the beach is above sea level the older it is. (*After R. I. Walcott.*)

average for the rate of uplift over the past few thousand years. We should notice that some areas of North America are currently falling with respect to sea level as a consequence of the last ice loading. But there is in fact nothing anomalous about this apparent paradox. The simple explanation is that when the ice load was depressing the crust in the Hudson Bay area, the mantle underneath was naturally shouldered aside to some extent to make room for the descending column under the continent. This displaced mantle material caused an upward bulge of the crust around the edge of the ice sheet, as we see in Fig. 3-15. Then when the ice melted, the depressed crustal area began to rise, but mantle material now began to flow in from the sides to fill any gap, and so the *forebulge* began to decay.

This sort of evidence from postglacial rebound that the earth is a mobile body influenced Alfred Wegener enormously, as we will see in the final chapter. Wegener reasoned that if such up and

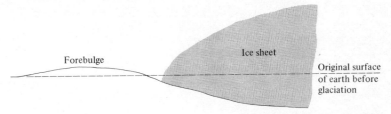

Figure 3-15 When the ice load depresses the continent, mantle material is pushed aside and naturally produces an *upwards* bulge of the crust around the rim of the icecap. When the ice melts, the depressed area begins to rise, whereas the forebulge falls.

down motions could occur so readily, then surely it is not too much to believe that horizontal motions of continents were also perfectly possible.

Chapter 4

How Old Is the Earth?

How old is the earth? When did it condense out of the disc surrounding the sun? How old are the mountains? How old is the earth's core? How long has the moon been circling the earth? How long has there been life on earth? And so on. Who among us has not asked himself or herself questions like these at least once? Before this century, it was impossible to supply reliable answers to any of these questions. As late as the 1890s Lord Kelvin's estimate of the age of the earth was over 100 times too small. Even by 1953, values calculated for the earth's age were about 25 percent too low.

The means for answering at least some of the above questions became available *in principle* in 1896 when Becquerel discovered radioactivity. In actual practice these means have been applied with high precision only in the past 20 years.

Let us look, then, at the phenomenon of radioactivity and see how it has come to be used as a remarkable natural clock.

RADIOACTIVITY

We know of 92 naturally occurring elements, like hydrogen, sodium, iron, etc. The vast majority of these are *stable*. That is, under the conditions found on earth they would live happily forever, just as they are now. A few exceptions, however, like uranium and thorium, are *unstable* and spontaneously change into other elements. Uranium and thorium slowly change into lead. Potassium slowly transforms into argon, while rubidium eventually becomes strontium. These transformations are accompanied by the emission of minute particles such as α particles, β particles, and γ rays. This act of emission is called *radioactivity*.

Now the critical aspect of radioactivity that interests us here is the fact that the radioactive transformations proceed at a *constant rate* (what precisely we mean by constant rate will be explained more carefully a little later). Each element, such as uranium, has its own specific rate, and what is extremely important, this decay rate cannot be altered significantly by terrestrial events. Hitting uranium with a hammer, for instance, or heating it, or dissolving it in acid does not change its rate of decay. Thus, radioactive elements "ticking away" in rocks, changing steadily into other elements provide us with the basis of a set of natural clocks.

To understand why the clocks involve the transmutation of elements and are so reliable, we have to examine radioactivity a little more deeply. Any *atom* is made up of a central *nucleus* of positive electric charge surrounded by a cloud of orbiting negatively charged electrons. In a neutral atom there are just enough negative electrons in the cloud to neutralize the positive nuclear charge. The chemical nature of the atom, whether it is identifiable as zinc, sulphur, mercury, or whatever, is fixed by the size of the positive charge on the nucleus. Thus hydrogen has the smallest nuclear charge and is said to have 1 unit. Potassium has 19 units of charge on its nucleus; uranium has 92. Now an alpha particle has 2 units of positive charge and is ejected from the nucleus of a radioactive atom. The nucleus of the remaining atom has, therefore, 2 fewer units of positive charge than had the original parent nucleus, and so the new *daughter nucleus* confers new chemical properties to the daughter atom. This explains how α emission produces a new element. The β particle carries 1 negative unit of charge, and so after a radioactive atom has emitted a β particle from its nucleus the charge on the resultant nucleus must be 1 higher than originally (since the loss of 1 unit of negative charge is equivalent to the gain of 1 positive unit.) Hence again, this time following β emission, we have an atom of a new element formed with a new nuclear charge. In contrast, γ rays are neutral; i.e., they carry no electric charge. Consequently, their emission from a nucleus does not change the nuclear charge and does not produce an atom of a new element. In fact, γ rays are a secondary radioactive phenomenon. By this we mean that they always follow some earlier event, such as the emission of an α or a β particle. They appear because an α or β emission sometimes leaves the new nucleus not surprisingly in an "excited" state. The nucleus is "unhappy" and radiates away some energy to become more stable. It is the γ ray which takes away this excitation energy. Some of these considerations are illustrated in Fig. 4-1.

The stability and regularity of the radioactive decay process is due to the fact that the transformation going on involves the nucleus. The nucleus is shielded very effectively from the hammer blows and corrosive acids of the outside world by the cloud of electrons composing the rest of the atom. Thus the moment at which a radioactive atom decays is not significantly influenced by natural terrestrial happenings, and the radioactive process runs on, neither helped nor hindered.

Half-Life

To understand how to tell the time with a radioactive clock, we will find it easiest to introduce a new unit of time—the *half-life*. Instead of telling time in hours or minutes, we will tell it in half-lives.

The half-life concept is simple. Suppose at any instant we have 1,000 million radioactive atoms of a given type. Then 1 half-life later there will be 500 million (i.e., one-half the original number) remaining. After another half-life has elapsed there will be 250 million left. After another half-life only 125 million will

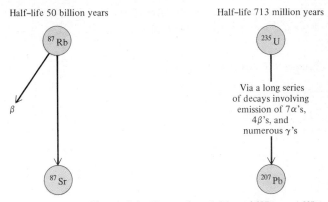

Figure 4-1   The radioactivities of ^{87}Rb and ^{235}U.

survive, and so on, the remaining number of radioactive atoms being halved every half-life. This is what we meant earlier when we spoke of radioactivity going on at a constant rate.

The important radioactive isotopes ^{235}U, ^{238}U, ^{40}K, and ^{87}Rb have quite different half-lives, ranging from 713 million years for ^{235}U to 50,000 million years for ^{87}Rb. We summarize them in Table 4-1.

How to Tell Time with a Radioactive Clock

We will begin an explanation by choosing a special case where it is especially easy to tell the time. Then we will go forward to see how things work in the general case. Let us suppose then that we wanted to measure the age of a rock which actually crystallized from a molten state exactly 713 million years ago and we propose to use the ^{235}U-^{207}Pb clock. Firstly, we will assume that when the rock crystallized, it trapped ^{235}U but no ^{207}Pb. Secondly, we assume that during the rock's 713-million-year history, it was sufficiently undisturbed that no ^{235}U or ^{207}Pb escaped from or entered the rock by diffusion. This is usually referred to as the *closed-system* assumption. We would then take this rock and analyse it chemically to determine how many ^{235}U and ^{207}Pb atoms it contains today. This would reveal that there are exactly as many ^{207}Pb atoms as ^{235}U atoms now present in the rock. But

Table 4-1 Values of Half-Lives of Useful Radioactive Elements

Isotope	Half-life in billions of years
^{40}K	1.31
^{87}Rb	50.0
^{238}U	4.51
^{235}U	0.713
^{232}Th	13.9

we would recall that each ^{207}Pb atom originally was a ^{235}U atom. Therefore, our measurement has told us that exactly one-half of the original ^{235}U atoms have changed into ^{207}Pb. However, this is exactly what would have occurred in 1 half-life of ^{235}U; therefore, we must conclude the rock has existed since crystallization for 1 ^{235}U half-life, i.e., from Table 4-1, 713 million years. This agrees with our original postulate.

In this illustration, we chose, as we promised, a particularly easy example where the rock was found to contain equal numbers of ^{235}U and ^{207}Pb atoms so that it was obvious that one-half the original ^{235}U had decayed to ^{207}Pb, and the age was therefore just as obviously 1 ^{235}U half-life. However, in general, the average rock one wished to date would not be 713 m.y. old; it would therefore not now contain equal numbers of ^{235}U and ^{207}Pb atoms; and the fraction of the original ^{235}U which had decayed since the rock crystallized would not be one-half. Let us suppose we had a rock which was well over 713 m.y. old and in fact, when analysed, was found to contain exactly twice as many ^{207}Pb atoms as ^{235}U atoms. Then, exactly following the previous argument, we would say evidently two-thirds of the original ^{235}U in this rock has decayed to ^{207}Pb. The rock must therefore have existed long enough for this to happen. To see exactly how long this was, we must now turn to Fig. 4-2. In this graph we have plotted vertically the fraction of any given initial number of radioactive atoms which have decayed and horizontally the number of half-lives during which such a decay has occurred. We have thus obtained a curve labelled *A*. To complete our age calculation, now we recall that exactly two-thirds of the original ^{235}U in the rock has decayed; so we find two-thirds, i.e., 0.67, on the vertical axis and draw a horizontal line to intersect curve A. We now drop a vertical line from this intersection point onto the horizontal axis and read off the corresponding number of half-lives, which in this case is 1.6. The rock has, therefore, existed for 1.6 ^{235}U half-lives, which is 1.6 × 713 m.y., i.e., 1,140 m.y. By exactly this process the ^{235}U-^{207}Pb age of a rock can always be found once the fraction of the original ^{235}U which has decayed has been found.

Figure 4-2 is very useful because it is not restricted to the ^{235}U-^{207}Pb scheme. K-Ar, Rb-Sr, ^{238}U-^{206}Pb, and ^{232}Th-^{208}Pb ages may all be found from it in the same way. Thus, in a K-Ar age determination, from the number of ^{40}Ar and ^{40}K atoms in a rock we can immediately calculate the fraction of original ^{40}K which has decayed. We then use Fig. 4-2 exactly as before to read off the corresponding number (n) of half-lives which must have elapsed since rock formation. But these are, of course, now ^{40}K half-lives; so the age of the sample is n × the half-life of ^{40}K, or n × 1,310 m.y.

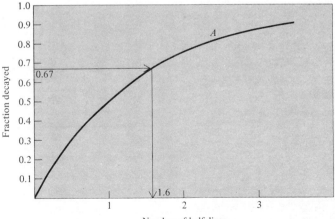

Figure 4-2 Graph of fraction of any given initial number of radioactive atoms which have decayed versus the number of half-lives during which the decay has occurred.

Now that we understand the principle of age determination, we can mention some of the difficulties experienced in practice. We mentioned earlier the closed-system assumption. This assumption is often violated. Many rocks have suffered some disturbance during their history, and therefore, one frequently finds ages which are too low because some of the daughter product has escaped by diffusion. Secondly, we have so far been assuming no daughter product was trapped in the rock at crystallization. This is usually a false assumption, but in the U-Pb, Th-Pb, and Rb-Sr techniques, straightforward ways of allowing accurately for this are well established. The presence of initially trapped ^{40}Ar in rocks is not, however, so easily detected but fortunately is often not important.

The good agreement which is found for a rock dated by the various techniques when the rock has not been seriously disturbed is shown in Table 4-2.

Before we go on to see the results which have been found with these various techniques, we notice that generally it is true to say that the older a rock the easier it is to date. This is simply because older rocks have accumulated more daughter atoms, and the daughter may, therefore, be more accurately measured. All the techniques work for rocks as young as 100 m.y. Below this age the K-Ar begins to outstrip the others in its effectiveness, and 99.9 percent of all ages (by U-Pb, Th-Pb, Rb-Sr, or K-Ar) less than 10 m.y. have been found by K-Ar dating. In special

Table 4-2   Reasonably Concordant U-Pb, K-Ar, and Rb-Sr Ages. *
(After Wetherill et al.)

Locality	^{206}Pb-^{238}U	^{207}Pb-^{235}U	K-Ar	Rb-Sr
Portland, Conn., U.S.A.	268	266	265	251
Glastonbury, Conn., U.S.A.	251	255	259	257
Spruce Pine, N.C., U.S.A.	370	375	349	352
Branchville, Conn., U.S.A.	367	365	382	
Parry Sound, Ont., Can.	994	993	970	
Cardiff Twp., Ont., Can.	1,020	1,020	1,000	970
Keystone, S. Dak., U.S.A.	1,580	1,600	1,520	1,570
Viking Lake, Sask., Can.	1,850	1,880	1,850	1,852
Bikita, S. Rhodesia	2,640	2,670	2,550	2,519

*Ages are in millions of years.

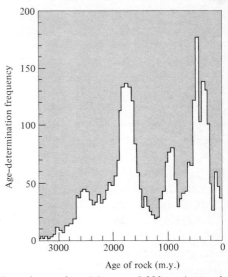

Figure 4-3 Distribution of ages found in over 3,000 analyses of continental rocks. (*After R. Dearnley.*)

circumstances, the K-Ar method may be applied to rocks as young as 10,000 years. Our clocks, therefore, contrast sharply with the ^{14}C clock, which is less accurate for older samples, cannot measure ages much greater than about 50,000 years, and is, therefore, generally more suited to archaeological and anthropological studies.

THE AGE OF THE EARTH

It is one thing to date a piece of rock chipped off a granite outcrop to calculate when the granite crystallized from the molten state. It is quite another to date a whole earth which has a radius of about 6,400 km, has over 70 percent of its surface covered by water, and has an accessible crust which is continually being heated, squeezed, and generally rejuvenated. The obvious approach would be to sample as much of the crust as is possible and date it with all the available means. The oldest reliable age obtained would then be the minimum age of the crust, and of course, the earth as a whole would be even older. The results of such an approach are illustrated in Fig. 4-3. In it are displayed the findings of over 3,400 age determinations on many continental materials. While there are a number of interesting points which may be made about this histogram, we are concerned here with only the one bearing on the earth's age—the point being that only a handful of these age determinations gave results greater than 3,000 m.y. or, as we shall often state it, 3 billion years (b.y.). Evidently, the vast preponderance of the earth's continental crust records the earth's history over the past 3 b.y. Only a few isolated older patches of crust have survived through the mountain-building episodes and general crustal activity of this period. But survive they did, in areas like the Ukraine, South Africa, and Antarctica, and in North America in Minnesota and Montana. Rocks from these areas have yielded ages variously in the range 3.1 to 3.5 b.y. Pride of place among these ancient terrains, however, belongs to a small area on the southwest coast of Greenland near the city of Godthaab. Between 1970 and 1972 a group of scientists at Oxford obtained convincing Rb-Sr ages of about 3.75 b.y. on numerous rock samples from this location. U-Pb data supported this value, while lower values were found with the K-Ar method because of loss of argon from the samples during thermal episodes in the area at some time following the original crystallization.

We must inevitably conclude that at least parts of the continental crust were originally formed over 3 b.y. ago, and at least one part has existed for over 3.75 b.y. We can say, therefore, with confidence, that *the earth as a whole is over 3.75 b.y. old.*

The question now arises: Just how *much* older than 3.75 b.y.? After all, the skin we call the crust could have formed long after the earth agglomerated into a huge sphere, and so even if the crust had lain undisturbed since its formation it could still significantly postdate the earth's formation.

What appears to be the correct answer to this question involves a somewhat oblique approach. We proceed in fact via the meteorites and see that they appear to have been formed about 4.6 b.y. ago, according to the Rb-Sr and K-Ar clocks. But considerations of lead-isotope evolution suggest the earth and the meteorites started life at the same time, and so we will conclude that the earth probably formed into a big sphere about 4.6 b.y.

ago. We will spend some time now, therefore, looking at the meteorites.

METEORITES AND THEIR AGES

Meteorites are fragments of rock and iron which have fallen on the earth from outside. Despite much investigation, their place and mode of origin are still unclear. Probably the most popular theory is that they come from the asteroidal belt of debris which orbits between Mars and Jupiter, looking like the remains of a disrupted planet. In fact, that there should be a planet at this distance from the sun was forecast from Bode's law (see Chap. 1). The fifth planetary orbit out was predicted to be 2.8 astronomical units from the sun [(24 + 4) ÷ 10 = 2.8, following the method described in Chap. 1]. At the time of the prediction no planet was known at this distance. In 1801, however, a small chunk of rock 450 miles across, named Ceres, was found in about the right place, to be followed over the years by the discovery of many other, though smaller, objects in the same general belt. Whether the remains really were ever once part of a planet, however, remains to be proved.

The meteorites fall into three categories: stones, stoney-irons, and irons. The stones are exactly what they sound like— chunks of rock composed of silicate minerals similar to ones found in terrestrial rocks. They also have small amounts of a nickel-iron alloy. There are two fundamentally different types of stoney meteorite—*chondrites* and *achondrites*. The former are characterized by the presence of small, roughly spherical aggregates, called *chondrules*, of the minerals olivine and pyroxene. The latter (i.e., the achondrites) do not have chondrules. Illustrations of these types are given in Fig. 4-4a and b. Irons (Fig. 4-4c) are pieces of nickel-iron alloy with only small amounts of silicate minerals. Lying conspicuously on the earth's surface, these objects provided early man with heaven-sent material for tools and weapon heads. Stoney-irons are roughly half and half nickel-iron and silicate.

Meteorites frequently have fiery passages through the earth's atmosphere because of the frictional heating, and large ones may

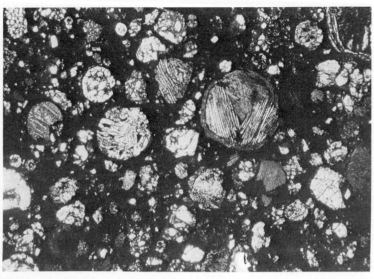

(a)

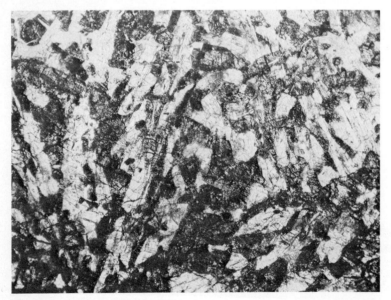

(b)

(c)

Figure 4-4 (a) A number of round chondrules are visible in the Tieschitz chondrite. The largest chondrule is about 1mm in diameter. (b) The texture of the Stannern achondrite displays dark patches of the mineral augite between light laths of felspar. Magnification is the same as in (a). (c) The Willamette iron meteorite weighs about 15 tons and is the fourth largest found. [(a) and (b) courtesy of J. A. Wood; (c) courtesy the American Museum of Natural History.]

explode noisily. Occasionally, devastating impacts may be produced, resulting in enormous craters such as the renowned Meteor Crater in north central Arizona, seen in Fig. 4-5. This feature is about 3/4 mile in diameter and 600 ft deep and is encircled by a rim varying in height between 100 and 200 ft above the plains. Shoemaker has estimated that the meteorite responsible weighed about 60,000 tons and struck the earth at about 10 miles/second. Fortunately such catastrophic arrivals are rare, on the human time scale.

The first Rb-Sr dates on meteorites were found in the 1956–1957 period by Schumacher, Herzog, and Pinson and Webster, Morgan, and Smales. Their results on chondrites indicated

Figure 4-5   The Meteor Crater in Arizona is 3/4 mile across and
600 feet deep. (*Courtesy the American Museum of Natural History.*)

that these meteorites were between 4.2 and 5.0 b.y. old. Many
subsequent analyses of gradually improving precision have sub-
stantiated these earlier results and show that almost all meteorites
were formed between 4.5 and 4.7 b.y. ago. The K-Ar clocks show
a considerably wider scatter of readings, but ones scattering
downwards from about 4.6 b.y., indicating original ages of 4.6 b.y.
but exhibiting the effects of loss of argon.

The U-Pb clock does not work in a straightforward way with
the meteorites. Something appears to have happened to the
uranium and lead concentrations so that reasonable ages cannot
be found. However, what is called the *lead-lead* clock still works,
and it is indeed via this chronometer that we make the transition
from the meteorites' age to that of the earth.

Let us suppose that when the meteorites first formed,
however they did and wherever they did, they all trapped within
their crystals the identical *kind* of lead, i.e., lead of a single
isotopic composition. Then this lead would be said to have the
primeval values of the lead-isotope ratios $^{208}Pb/^{204}Pb$, $^{207}Pb/$
^{204}Pb, and $^{206}Pb/^{204}Pb$. Now the ^{204}Pb isotope is not produced by
any natural radioactivity and so does not vary in amount in the

meteorites with time (unless it is removed or added by some physical-chemical process). The concentrations of the isotopes ^{208}Pb, ^{207}Pb, and ^{206}Pb are, however, continually increasing because of their production by the radioactive decays of ^{232}Th, ^{235}U, and ^{238}U respectively. So the ratios $^{208}Pb/^{204}Pb$, $^{207}Pb/^{204}Pb$, and $^{206}Pb/^{204}Pb$ continually increase with time. And, in fact, if we plot a graph of how the last two ratios vary together in time in a single meteorite, we get the curve shown in Fig. 4-6a. This *lead growth curve* naturally begins at the point labelled

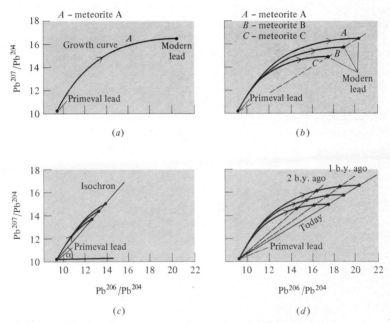

Figure 4-6 (a) Lead growth curve of a single meteorite. When the meteorite formed, it trapped primeval lead. Subsequently, because of the radioactive decay of uranium, the lead composition grew along the curve to the modern lead it is today. (b) Different meteorites trap different amounts of uranium relative to lead, and so each meteorite has its own lead growth curve. (c) At any time after formation, the lead compositions for all meteorites will lie along a straight line called an isochron, provided all meteorites formed simultaneously and with the same primeval lead isotope composition. (d) As the isochron rotates about a pivot at the primeval lead point, the points of intersection of this line with the growth curves give the lead isotope compositions of the meteorites at later and later times.

primeval lead and ends at *modern lead,* meaning the lead now found in this particular meteorite. We must expect, though, that no two meteorites will trap the same amount of uranium relative to lead during formation, and therefore, we expect that each meteorite will have its own *lead growth curve, as we illustrate in Fig. 4-6b.* At this stage it might seem that we have a rather chaotic situation to handle, with all these different growth curves being followed at different rates. However, there is a saving grace. At any single instant after meteorite formation, the points on the growth curves representing the instantaneous lead-isotope compositions in the various bodies will *all* fall on a single straight line, as shown in Fig. 4-6c. Furthermore, this straight line goes through the original primeval lead point. The gradual movement of the meteorite lead-isotope composition points along the growth curves is, therefore, given by the movements of the points of intersection of the straight line with the growth curves as the straight line swings in an arc about a pivot fixed at the primeval lead point. This is made clear in Fig. 4-6d. The straight line is called an *isochron,* which merely means *same time.* Our description of this lead-lead clock is complete when we note that the time

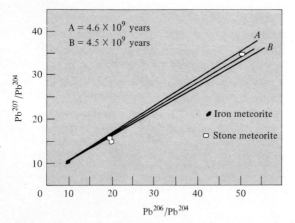

Figure 4-7 In 1956, C. Patterson found that lead samples from iron and stone meteorites did lie on a straight line, that is, they defined an isochron. The angle made by this line with the horizontal axis showed that these meteorites were about 4.6 b.y. old.

which has elapsed between formation of the meteorites and their arrival at a particular isochron position is calculated from the angle which the isochron makes with the direction of the horizontal axis, i.e., angle α in Fig. 4-6c. We have now reached the moment of truth where we must face the test of whether real meteorites do have lead-isotope compositions which fall on an isochron in a $^{207}Pb/^{204}Pb$ versus a $^{206}Pb/^{204}Pb$ plot. This moment was first experienced in 1956 by C. Patterson, who found that indeed the lead from three stone meteorites and two irons which he had analysed had isotope ratios which fell on an isochron whose angle to the horizontal showed that it was about 4.6 b.y. since all the meteorites analysed had shared a common primeval lead. Patterson's historic graph is shown in Fig. 4-7. The iron-meteorite data points fall at the extreme left-hand end of the isochron because they contain virtually no uranium, and therefore, the lead in them is essentially the primeval lead captured by the meteorites 4.6 b.y. ago.

Thus the Rb-Sr, K-Ar, and lead-lead clocks all agree that the meteorites evidently formed about 4.6 b.y. ago. Or if they are older than this, they were so severely metamorphosed 4.6 b.y. ago that no trace of an earlier history remains.

Now we ask, what would we find if we took the leads from recent terrestrial sediments and volcanoes and plotted them on Patterson's lead-lead graph? The answer is we find that such lead compositions fall in a group on and about the meteorite isochron (Fig. 4-8). When Patterson discovered this, he concluded that the earth too began life 4.6 b.y. ago and had at that time the same kind of primeval lead as did the meteorites. Later measurements on meteoritic lead have confirmed the reality of the meteorite isochron, and the value of 4.6 b.y. is now widely quoted as "the age of the earth."

While it is probable that the earth is about 4.6 b.y. old, we should add just a few qualifying comments about the above arguments. The simple reading of the lead-lead clock assumes that the various samples of lead analysed all developed in their own local closed systems with particular U/Pb ratios. Each sample, therefore, developed along its own growth line to the present. While this requirement is probably effectively met in the

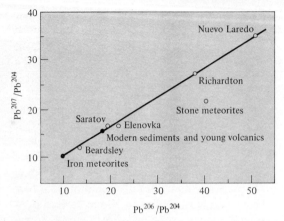

Figure 4-8 Later work added more points to the 4.6 b.y. meteorite isochron. When the *earth's* modern lead is entered on such a diagram, we find it falls on or near the meteorite isochron and so it is usually concluded that the earth as well as the meteorites formed 4.6 by.y. ago. (*After Murthy and Patterson.*)

meteorites (i.e., assuming the earlier mentioned disturbance of their U/Pb ratios is too recent to affect the lead-lead clock), it is certainly not met in the terrestrial samples. Lead now found in modern sediments and volcanic rocks must have had a more complex history, and therefore, it is somewhat surprising that modern terrestrial lead lies so close to the meteorite isochron. This puzzling point remains to be resolved.

In summary we may say then that the earth is over 3.75 b.y. old since it must be older than the oldest crustal rocks so far found. From lead-lead dating of meteorites and modern terrestrial lead we get fairly strong evidence that the earth formed about 4.6 b.y. ago at the same time as the meteorites (which *are* well dated), and as a working hypothesis, therefore, we take 4.6 b.y. to be its age.

EVOLUTION OF THE CORE AND MANTLE

We have just seen that the earth probably formed into its present size 4.6 b.y. ago. We do not know, however, the physical state it was in at that time. Throughout the nineteenth century and the first half of the present century, it was generally assumed that the earth was originally molten. It was believed that when the originally diffuse earth material began to condense into a larger and larger ball, the energy released on impact by the infalling matter would be enough to cause the earth to melt. Now, it is true that there is quite enough energy released in this way to melt the whole earth *if the energy is effectively used for melting.* However, the growing, heating sphere will be radiating heat away into space continuously; so whether the earth is molten at the end of its formation depends critically on the balance struck between gravitational accretional heating and radiative cooling. Unfortunately no adequate calculation of this balance has been made. Another reason for supposing the earth was hot initially is that one expects that 4.6 b.y. ago there would have existed a number of relatively short-lived radioactivities, such as ^{26}Al, which are no longer found on earth. Since radioactivity generates heat, these activities, which died away long ago, could have added a significant amount of heat to the earth in its infancy. Again, however, this is a difficult factor to evaluate.

Now, in Chap. 2 we saw that seismology has revealed that the earth has a core and a crust. If the earth were molten 4.6 b.y. ago, then almost certainly this was when the differentiation of the earth material occurred, and the core and the crust are essentially as old as the earth.

During the past 30 years a competing theory of initial state has been developed wherein it is argued that in the beginning the earth was relatively cool throughout (this could mean as much as 1000°C) and was not molten. Therefore, 4.6 b.y. ago the earth was of a uniform composition with no differentiation into a core, a mantle and a crust having occurred. Urey has argued that in the subsequent billions of years, the buried radioactivity gradually raised the temperature inside the earth until in certain volumes the temperature exceeded the melting point of iron, but not of the surrounding silicates since it is more difficult to melt them. Small pools of iron are then envisaged as having collected and having "seeped" inwards towards the centre of the earth because of the high density of iron. As a result a metallic core gradually grew until it reached its present size. This theory, which is so different from the hot-origin concept, was very popular during the 1950s and 1960s but has lost some support in recent years. If it seems surprising that two such contrasting theories of the earth's beginning have on average enjoyed equally strong support over the years, it should be realized that what we are doing is equivalent to trying to solve a murder mystery in which the crime was committed 4.6 b.y. ago, to which there was no eyewitness, and about which we have only a small amount of remaining circumstantial evidence.

EVOLUTION OF THE CONTINENTAL CRUST

While it is difficult to speak with confidence of the earth's beginning and the development of its core and mantle, somewhat more may be said about the evolution of the crust. We know from the study of the southwest Greenland rocks that at least some of the continental rocks are 3.75 b.y. old. And, from the histogram in Fig. 4-3, we can see that much of the continental crust has been in existence at least 2.5 b.y. For instance, in Fig. 4-9 we see the known extent of continental North American rocks 2.5 b.y. ago. Not all these rocks are now exposed at the surface since they are frequently covered by younger sedimentary rocks, but their existence at depth has been proved by the Rb-Sr and K-Ar dating of drill cores. However, it is still not clear what fraction of the present continental crust existed 3.75 b.y. ago, and has been rejuvenated by mountain-building episodes, and what fraction has been added by differentiation from the interior of the earth with time.

Detailed geological mapping of all the accessible continental areas carried out over many years has revealed that mountain-building episodes have recurred throughout the history of the continental crust. During any one of these there is usually something like the following sequence of events: (1) A long segment of the continent is depressed into a trough in which sediments and volcanic rocks accumulate; (2) the deeper rocks are metamorphosed, and there is intense folding of the rocks, intrusion of granitic rocks, and volcanic activity; (3) the trough, usually called a *geosyncline*, widens and (1) and (2) are repeated; (4) eventually there is a general uplift of much of the belt.

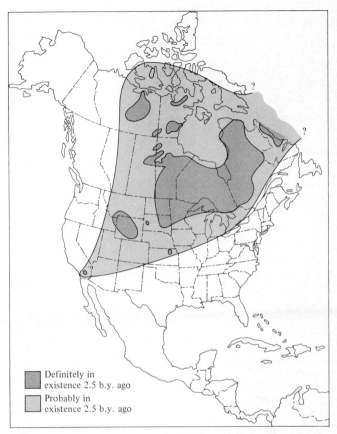

Definitely in
existence 2.5 b.y. ago
Probably in
existence 2.5 b.y. ago

Figure 4-9 The known extent of North American continental rocks
2.5 b.y. ago. (*After W. R. Muehlberger, R. E. Denison, and E. G. Lidiak.*)

Conventional mountain topography is then produced by erosion
of the elevated rocks. Over hundreds of millions of years the
visible mountains are eroded until no elevated topography remains. Yet the peaks are not simply removed as though skimmed
from the earth with a knife. As we saw in Chap. 3, the continental
mountains are floating on the denser underlying mantle like logs
in water, and so, as the material of the mountains is carried away

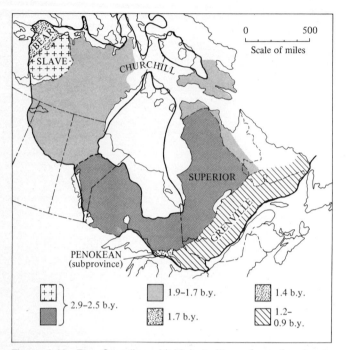

Figure 4-10 The Canadian shield is built up from the deeply eroded roots of ancient mountain chains. The dates indicate roughly when these mountains were built. (*After C. H. Stockwell.*)

by rivers and streams, the original deep mountain roots (see Fig. 3-10) are no longer necessary. The mountain column, therefore, will slowly rise to maintain isostatic equilibrium. Thus in completely eroding an uplifted block, far more than the initial obvious amount of elevated material must be removed. When the mountains have essentially been removed, what is then left exposed at the earth's surface are the rocks that were the roots of the original mountains. Such ancient, deeply eroded, light but hard and brittle, granitic and gneissic rocks form the *Precambrian shield* areas of all the continents. In Fig. 4-10 we show how the Canadian Shield is made up of these relicts of long-gone mountain chains. The oldest such belt, which we might call the heart of Canada, is approximately 2.6 b.y. old and is known as the

Superior province. To the northwest of this lies the Churchill province, about 1.8 b.y. old; and to the southeast the billion-year-old Grenville province stretches southwest and northeast to the Atlantic Ocean. Outside the shield area the continent is coated with younger sedimentary rocks which make the determination of the full extent of the old mountain roots difficult.

The history of all the continents may then be said to be one of repeated mountain building, followed by planing off of the elevated areas. These actions have gone on unabated for 3.8 b.y. until the present. The forces responsible for this behaviour at the earth's surface and the details of this behaviour during the past 200 million years will form the basis of the final chapter of this book. We have also been discussing so far strictly the continental crust. The oceanic crust has experienced a totally different history, yet one which has been intimately linked with that of the continents. In the final chapter we will see how these two histories may be synthesized into a whole crustal evolution. Now we will turn from the *physical* development of the earth to its *biological* evolution.

BIOLOGICAL EVOLUTION

The study of fossils (paleontology) from sedimentary rocks has shown that life has existed on the earth for over 3 b.y. An adequate documentation of it, however, covers only the last 600 m.y. Prior to this time, relatively insignificant numbers of organisms had hard phosphatic or calcium-carbonate components, which are the prerequisite for a good fossil afterlife. These last 600 m.y. are known as *Phanerozoic time*, and this was long ago subdivided, as shown in Table 4-3, by geologists as a result of painstaking stratigraphic and paleontological analyses of rocks and fossils from around the earth. The Phanerozoic is divided into three *eras*: Paleozoic, Mesozoic, and Cenozoic, which mean *ancient life, mediaeval life,* and *present life* respectively. The eras themselves are subdivided into periods with names like Cambrian, Permian, and Tertiary, which are shown in Table 4-3 in chronological order. The actual numerical estimates of times of occurrence of these periods were, of course, made with the aid of

Table 4-3   Diary of Planet Earth

Millions of years before present	Era	Period	Epoch
0—	Cenozoic	Quaternary	Recent
3_			Pleistocene
7_		Tertiary	Pliocene
25—			Miocene
40—			Oligocene
55—			Eocene
			Paleocene
65	Mesozoic	Cretaceous	
135—		Jurassic	
195—		Triassic	
225	Paleozoic	Permian	
280—		Pennsylvanian (Carboniferous) Mississippian	
350—		Devonian	
400—		Silurian	
440—		Ordovician	
500—		Cambrian	
600	Precambrian		
3,300—			
3,700—			
4,600—			

Physical events	Biological events
	Modern man
Ice Age	Early man
Cascadian mtn building	Hominids
	Ancestral dogs & cats
Alpine mtn building	Ancestral apes
	Ancestral horses, cattle, & elephants
	First primates
Laramide mtn building (Rocky Mountains)	Extinction of dinosaurs
	First flowering plants
Nevadan mtn building	
	First birds
	Peak of dinosaurs
	First dinosaurs
	First mammals
Appalachian mtn building	Extinction of many Paleozoic species
	Rise of reptiles
	Coal-forming forests
Hercynian mtn building	
Acadian mtn building	First reptiles
	First amphibians
	First insects
	First trees
	First air-breathing animals
Caledonian mtn building	First land plants
	First vertebrates (fish)
Taconic mtn building	First corals
	Earliest abundant fossils (trilobites & graptolites)
	Scanty fossil record
≈	
	Algal & bacterial remains
	Oldest rocks (Greenland)
	Origin of earth

Figure 4-11 A prominent life-form in the Cambrian era was the trilobite. (*Courtesy the Royal Ontario Museum.*)

K-Ar, Rb-Sr, and U-Pb dating techniques applied to rocks which could be correlated with the time scale.

From the Cambrian period, with the first abundant fossils, such as the trilobite in Fig. 4-11, life gradually evolved in complexity. The earliest fish-like vertebrates appeared in the Ordovician; trees, spiders, and winged insects in the Devonian; coal forests and reptiles in the Carboniferous; dinosaurs and mammals in the Triassic; birds in the Jurassic; flowering plants in the Cretaceous. In the Tertiary period came ancestral horses, cattle, elephants, dogs, and apes. In Fig. 4-12 we see how the line

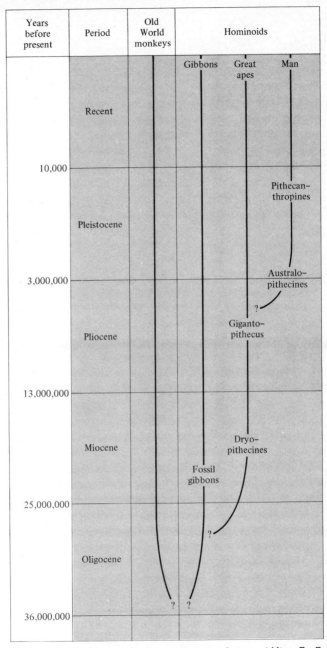

Figure 4-12 The stages in the evolution of man. (*After R. B. Eckhardt.*)

leading to the Old World monkeys split off from the hominoid lines in the early Oligocene, perhaps 38 m.y. ago. A little later the Dryopithecines split off from the line leading to the modern gibbons. Finally, at some uncertain time in the Pliocene, the hominid line, which culminated in the human race, diverged from the line which yielded the great apes.

Creatures even remotely resembling modern man have evidently been on earth for much less than 4 m.y. Since the earth is over 4,000 m.y. old, we conclude that human beings have existed for considerably less than 1/10 of 1 percent of the earth's history. If we consider that life has been abundant and varied during only Phanerozoic time, then we see that this has occupied about 13 percent of the earth's existence. It would be wrong, however, to conclude from these figures that life on earth is a relatively recent phenomenon. Traces of it have been found from time to time in Precambrian rocks in various part of the world. Bacteria have been recorded in the Fig Tree Series sediments of Swaziland (see Fig. 4-13), which dating indicates may be 3.5 b.y. old; algal structures have also been found in the Bulawayo Limestone of Rhodesia, which is probably 3 b.y. old. A variety of microfossils is known from the Gun Flint Formation of southern Ontario, which is about 1.9 b.y. old. Not until the latest Precambrian, however, is evidence for animal remains very clear. The best documented example occurs in the Pound Quartzite of Ediacara in South Australia, where fossils of jellyfish and worms have been found.

EVOLUTIONARY EXPLOSIONS AND EXTINCTIONS

It has long been known that the numbers and forms of living creatures did not simply increase in a steady fashion throughout Phanerozoic time. The fossil record is punctuated by dramatic

Figure 4-13 Photographs taken with an electron microscope of fossil bacteria in the Fig Tree Sediments of South Africa. These are the most ancient fossils known—over 3 billion years old. The short line in the photographs is one millionth of a metre (1μ) long. (*Courtesy of L. S. Barghoorn.*)

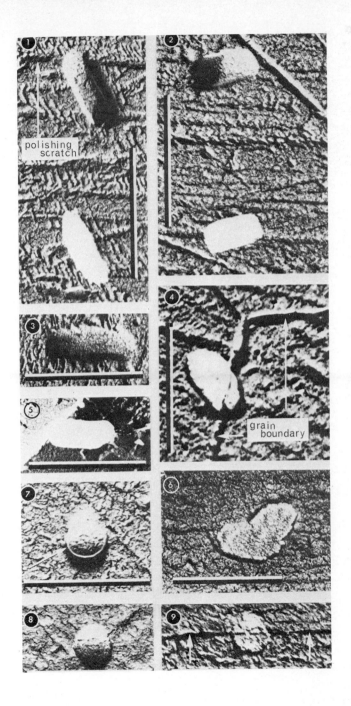

proliferations and extinctions. In 1963 Uffen proposed that the extinctions resulted from the effects on life of supposed increases in cosmic irradiation during times when the earth's magnetic field reversed. In our chapter on magnetism we will look into this possibility. It suffices here to say the question of a correlation of extinctions with field reversals remains an open one and that others have suggested these evolutionary-rate variations result from sea-level changes, climatic changes, and variations in the composition of the atmosphere among numerous other possibilities.

Chapter 5

Magnetism and the Earth

The fact that some rocks behave as magnets was known to the ancient Greeks, while the magnetic compass was used by the Chinese in the eleventh century. The Chinese were probably the first people to appreciate that the earth has a magnetic field. The modern study of the earth's field dates from the time of Queen Elizabeth I of England, whose physician, William Gilbert, wrote the first book on magnetism, *De Magnete*. Gilbert realized that the magnetic field varied over the surface of the earth as though the earth's body were uniformly magnetized throughout. Newton's friend Edmund Halley observed as early as the seventeenth century that some features of the terrestrial field tend to drift westwards. Progress towards the understanding of the origin of the field was extremely slow, however, and only during the last 30 years has an outline been sketched by Elsasser and Bullard of the probable cause.

We will begin by outlining the observed characteristics of the terrestrial magnetic field, including both its present and past

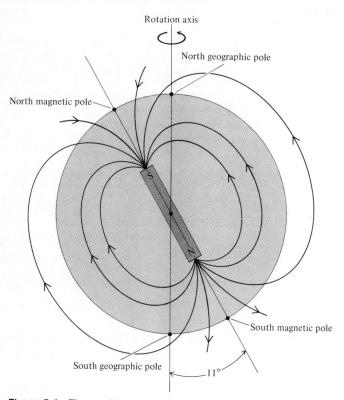

Figure 5-1 The earth's magnetic field is very similar to that which would be produced by a bar magnet at the centre aligned 11° from the rotation axis.

behaviour. Finally we will consider the various theories of its origin, giving particular attention to the *self-generating dynamo theory*, which appears to be the only likely mechanism for generating the observed field.

CHARACTERISTICS OF THE EARTH'S MAGNETIC FIELD

To a very good approximation, the earth's magnetic field may be represented by that of a bar magnet, or dipole, located at the centre of the earth, the axis of the dipole being displaced roughly 11° from the rotation axis (see Fig. 5-1). The magnetic poles are in fact defined as those points on the surface of the earth which would be intersected by the continuation of the axis of the best-fitting dipole.

The magnetic field is a vector quantity, having both a magnitude and a direction associated with it. The customary parameters usually given with the field are shown in Fig. 5-2. The two chief angles of interest are those of *declination* and *inclination*. Declination measures the angular deviation between geographic north and the direction shown on the compass needle at the point in question and has long been of importance in navigation. Inclination is the angle which a freely swinging magnetic needle makes with the horizontal. Declination angles east of north are called *positive,* as are inclinations below the horizontal. Intensities most quoted are the horizontal component value and the total intensity. For several centuries at least some of these parameters have been measured continuously at a number of observatories, and some of the earliest results have been very interesting. However, detailed recording of the terrestrial field at many points has been carried out during only the

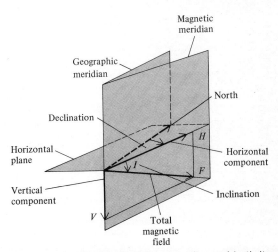

Figure 5-2 The magnetic field is a vector quantity, and both its magnitude and direction must be specified.

second half of the nineteenth century and throughout this century. The results of this monitoring have been most fruitful, and we will now consider several of these in some detail.

INTERNAL ORIGIN OF THE FIELD

In a classic work in 1839, the great German mathematician Gauss performed a spherical harmonic analysis of the earth's field as it was then known from observatory values. The most important feature of such an analysis is that it is possible to say that such and such a harmonic is due to an external source and such and such to an internal source. The immediate result of this analysis was that Gauss showed that essentially all the field was of internal origin. Since then, with the proliferation of observatories and the vast increase in good data, new spherical harmonic analyses have confirmed that the vast majority of the surface field is internally produced while some fluctuations are produced by external effects.

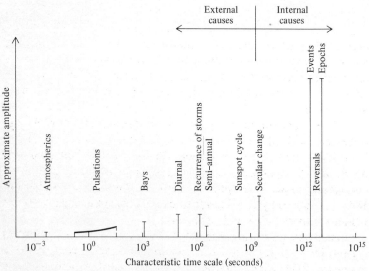

Figure 5-3 The range of geomagnetic field fluctuations. (*After G. D. Garland.*)

VARIABILITY OF THE FIELD

One of the most striking features of the terrestrial field is its variability with time. The field ranges spatially in value from about 70,000 gammas (γ) to 20,000 γ from the poles to the equator, but there are temporal fluctuations from these values of various magnitudes having periods from less than a second up to millions of years. The range and type of fluctuation are outlined in Fig. 5-3. Of these we will discuss the daily, or diurnal, variations, the secular change, and the reversals of the field.

Daily Variation of the Field

Examination of the daily records at a magnetic observatory reveals a rough daily trend to the magnetic fields, which is repeated daily on days not showing higher than normal variations. Similar variations are seen at different observatories located at the same latitude, provided local time is used in each case in looking for the trend. The latter point clearly indicates that the sun is the chief influence behind the trend. Typical daily trends in the magnetic parameters as a function of latitude are shown in Fig. 5-4. Largest field-strength variations are seen in equatorial regions, while declination variations are greater at higher latitudes. While these variations are very easily measured, they are not major disturbances of the field. The equatorial field variations referred to represent fluctuations in the total field of only about 0.5 percent, or less, and the daily variations leave no lasting effect on the field.

More detailed analysis of the magnetograms reveals the presence of both the 24-hour solar-induced variation and a lunar-controlled 25-hour variation, the two usually being denoted by S_q and L_q, respectively. Solar and lunar tidal motions in the conducting upper atmosphere, originally proposed by Stewart in 1882, are generally considered to cause a dynamo type of effect which yields the S_q and L_q variations.

The basic daily variation is usually disturbed by irregularities caused by particles fired from the sun. This *solar wind* carries a magnetic field which interacts with the earth's field, causing measurable fluctuations at the earth's surface. *Magnetic storms*

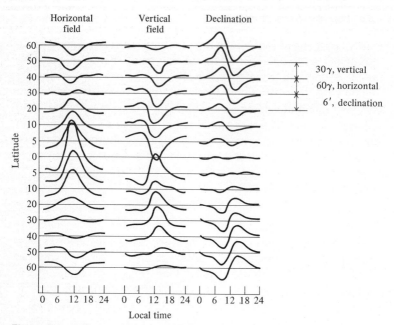

Figure 5-4 Daily trends in magnetic parameters as a function of latitude. (*After Matsushita.*)

show indisputable correlations with sunspot activity. The onset of such storms is usually marked by a small increase in the horizontal field intensity (*H*), following which there is a deep negative excursion of hundreds of gammas in a few hours. Recovery to normal field levels takes about 2 days.

The solar wind is now known to exert a major influence on the overall shape of the earth's field as seen from afar. This may be seen in Fig. 5-5, where the solar wind sweeps the lines of force of the earth's field into a long tail away from the sun.

When charged particles from outside enter the earth's field, they move in spirals along its lines of force (see Fig. 5-6). As the particles spiral into the polar regions of higher field strength, successive spirals get closer together until a point (a magnetic mirror) is reached where the forward motion ceases and the particles spiral back along the line of force. In a stable situation, charged particles would thus oscillate backwards and forwards, spiralling from one polar region back to the other along lines of force of the earth's field. Belts of charged particles thus trapped, including cosmic rays not from the sun, were first detected by the artificial satellite measurements of Van Allen in 1958 and are known as the *Van Allen radiation belts* (see Fig. 5-5). Changes in solar wind strength can upset the stability of these belts, causing charged particles to be precipitated into the atmosphere of the earth where collision effects produce the beautiful auroral displays well known to the inhabitants of northern latitudes.

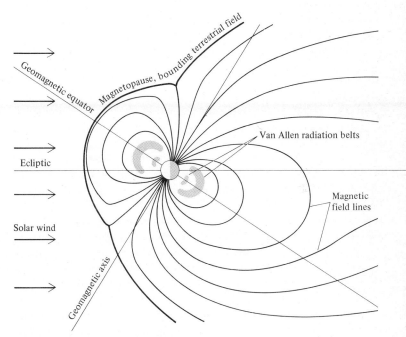

Figure 5-5 The stream of radiation from the sun, the solar wind, "blows" the lines of force of the earth's magnetic field into a tail pointing away from the sun. (*After L. J. Cahill, Jr.*)

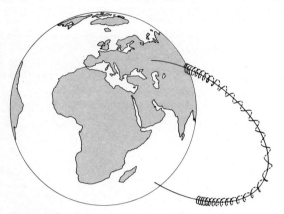

Figure 5-6 In the Van Allen radiation belts of Fig. 5-5, charged particles spiral backwards and forwards as shown here about lines of force of the earth's magnetic field. (*After J. A. Van Allen.*)

Secular Variation of the Field

Magnetic records kept since the sixteenth century clearly show that the earth's field undergoes considerable variations at any point both in intensity and direction with periods of the order of 100 years (Fig. 5-7). Variations in intensity of over 100 γ /year occur in some locations, persisting for many years. Changes of about 20 percent in field intensity may occur in some localities over several decades, indicating that this "secular" variation has an internal origin.

If world maps of such secular variation are prepared, it is apparent that some features on one map are displaced westwards from their positions on maps of earlier epochs. This drift of certain features of the secular variation is an aspect of what is known as the *westward drift*. Bullard and others (1950) analysed such data and found that the rate of drift varied with latitude and that an average rate of westward drift for all latitudes was about 0.18° per year for features of the nondipole field. If such a rate were continued, in time it would lead to a complete rotation of the nondipole features through 360° in about 2,000 years. Thus the earth's field in an area when averaged over several thousand years would be even more nearly that of a central dipole than it is now.

Reversals of the Field

The variations in the earth's field we have been considering until now have been of the order of 1 percent or less in the case of daily changes and as much as 20 percent over several decades in the case of secular change. When we go to a time scale involving millions of years, we discover much more dramatic changes. We find that the earth's field has totally reversed itself in direction many times. During the act of reversal the intensity of the field evidently falls to extremely low values (perhaps 1/10 normal) before recovering roughly its original value in the opposite direction.

In the first part of this century, several workers such as Matuyama, Brunhes, and Chevallier realized that such past behaviour of the earth's field was a distinct possibility. This realization was based upon results they obtained in the study of the magnetization of rocks. We will therefore discuss rock

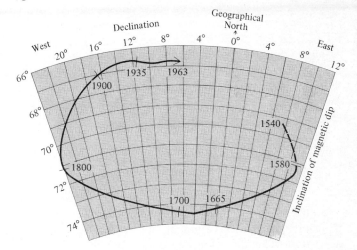

Figure 5-7 Magnetic field variation near London over the past few hundred years. (*After A. Holmes.*)

magnetism briefly to set the scene for the modern developments in the study of field reversals, which has been at the heart of the revolution in understanding in the 1960s of the earth's surface behaviour.

When the molten lava from a volcano cools down to become a solid at normal temperatures, it is found to be magnetized in the direction of the earth's magnetic field at that point. The rock is thus like a compass needle which has been frozen to preserve the direction of the earth's field. Many studies in the past 20 years have in fact shown that this compass reading is accurately frozen in, certainly for tens of millions of years and often for hundreds of millions of years. The reading of course may well be disturbed by heating or squeezing of the rock during mountain-building processes and by a variety of other mechanisms. Very often however, these effects can be allowed for and the original reading safely recovered. Studies of this nature fall under the heading of *paleomagnetism*, unquestionably one of the most fascinating fields in earth science.

The rock type which has been studied in by far the greatest detail in paleomagnetism is basalt, a dense, dark-looking volcanic rock erupted from volcanoes in Hawaii, Iceland, mid-ocean ridges, and many continental areas. The magnetism of the rock resides principally in an iron oxide called *magnetite* (Fe_3O_4). Indeed, paleomagnetism is really based on the magnetism of the three oxides—magnetite, hematite (Fe_2O_3), and ilmenite ($FeTiO_3$). Magnetization acquired by cooling in this way from a high temperature in the earth's field is known as *thermoremanent magnetism*, or TRM. Another chief way in which rocks acquire magnetization is known as *depositional remanent magnetization* (DRM). This process occurs during the formation of sedimentary rocks. When already-magnetized rock grains washed from continental areas are deposited at the bottom of seas, they will fall like tiny freely swinging compass needles and will tend to align themselves in the direction of the earth's field at the point of deposition. This direction is then frozen in by compaction of the sediment into rock. Sedimentary rocks are also often magnetized by *chemical remanent magnetization* (CRM) in which already-magnetized crystals change chemically and are remagnetized or

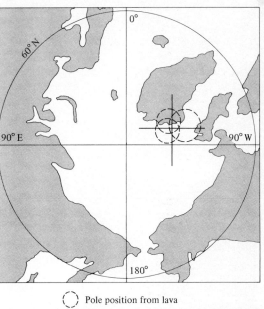

( ) Pole position from lava
+   Pole position according
    to observatory

Figure 5-8 Cox and Doell's test of paleomagnetic technique with modern lavas. (*After R. Doell.*)

in which new minerals grow after deposition and are magnetized.

The first test of the principles of paleomagnetism is obviously to take lavas erupted in historically recent times from places where the magnetic field was known from local observatory measurements and measure the direction of the magnetization in these rocks. If the magnetic direction measured in the rocks does not coincide within reasonable error with that measured for the field at the observatory at the time of eruption, obviously the method can hardly be expected to work when rocks millions of years old are used. The positive results of such a test carried out on Hawaiian volcanic rocks are shown in Fig. 5-8. From the directions of magnetization recorded in three lavas erupted in 1907, 1935, and 1955, Cox and Doell calculated that the North Magnetic Pole was located with a 95 percent probability within

Figure 5-9 Many natural exposures of lavas show reversals in direction of magnetization (indicated by the arrows) as one goes up the pile.

the three small circles shown. The cross indicates where the North Geomagnetic Pole was according to data from the local magnetic observatory established at Honolulu around 1907. Fortunately for the method of paleomagnetism, the agreement between the magnetic record written in the rocks and that in the observatory is complete within the experimental errors involved. These results and many others have clearly shown that, at least for a few tens of years, the basaltic volcanic rocks preserve an accurate memory of the earth's magnetic field from which we can calculate the position of the earth's magnetic poles at the times when the rocks crystallized.

When volcanic rocks ranging in age from recent to 15 million years old are analysed, they indicate geomagnetic pole positions clustered around the present *geographic* poles (i.e., the poles at which the earth rotates) rather than about the present geomagnetic poles, which we have seen are displaced significantly from the earth's rotation axis (Fig. 5-1). A number of important deductions may be made from this. Firstly, the consistency of the data indicates that the magnetic record is preserved well for at least 15 million years in the rocks. Secondly, the grouping around the geomagnetic poles indicates that the axis of the earth's magnetic dipole probably moves around the rotation axis during the course of about a million years. Furthermore, the results are clearly consistent with the earth's field having been that of a dipole for at least the last 15 million years.

This then is a glimpse of the fundamentals of paleomagnetism, after which we can go forward to the study of reversals of the earth's magnetic field. Just before doing so, however, we note that many important tests have been devised to make sure that the rock-magnetic record is being read correctly. Misleading secondary components of magnetization are frequently present in rocks. Thus lavas can acquire a *viscous remanent magnetization* (VRM) merely by lying around for years in the earth's field long after they have cooled. Lightning discharges also produce spurious magnetic directions. Considerable success, however, has been achieved with "magnetic washing" techniques which remove such disturbances. The details of such processes are given in a reference at the end of this book (Strangway, 1970).

What the early paleomagnetic workers had essentially found, and what has been amply confirmed in modern studies, was that when one examines a pile of many horizontal lava flows, some of them are magnetized in one direction while the rest are magnetized in exactly the opposite direction (see Fig. 5-9). This irregular alternation of polarity as one goes up a pile of lavas has only two straightforward possible explanations. Firstly, as did the early workers, one might guess that the earth's magnetic field reverses its direction every so often and the lavas capture a record of this. Or one might suspect that the earth's field has not reversed at all but that certain lavas misbehave as compass needles and for some reason or other point in exactly the opposite direction from the conventional compass and the "conventional lava." For a number of years, in fact, these two conflicting views were hotly disputed. Recently, however, it has become quite clear that the earth's magnetic field has reversed many times in the past, at least 100 times and probably many, many more. This is so despite the fact that the odd volcanic rock has been found which has the contrary properties mentioned above and points its magnetic finger in exactly the wrong direction.

The controversy was resolved in a beautifully clear-cut way by a combined attack using paleomagnetic techniques and the potassium-argon method of dating which we described earlier. The analyses were carried out essentially by four scientists, Cox, Doell, and Dalrymple in the United States and McDougall in Australia. In principle the following argument was used: If the earth's magnetic field does reverse occasionally, then the history of these reversals should be identical at all points of the earth's surface because the earth's field is a worldwide phenomenon. If, however, no field reversals have occurred and the curious magnetic direction reversals in the lavas are due to peculiarities of the lavas, then one would not expect any correlation in age between reversely magnetized lavas in Iceland, say, and Australia. This argument was tested by the K-Ar dating of normally and reversely magnetized lavas from such scattered places as Hawaii, Iceland, and Africa. Samples of the lavas were collected with portable drills, analysed in the laboratory for their magnetic directions, and then used in K-Ar analyses. The results were astonishingly clear. Normally magnetized rocks formed one set of age groups, while reversely magnetized rocks fell into other groups. Thus it was clear that all rocks of a given age were all of one magnetic direction. Such a worldwide result could be explained only by changes in something like the magnetic field of the whole earth, and so the reversals of the earth's field were proved to exist. A time scale of these reversals determined by paleomagnetic and K-Ar studies is shown in Fig. 5-10.

There are several notable things about this field-reversal time scale. Firstly, it spans a time interval from the present back about only 4 million years. This may seem like a long time, but it is in fact barely one-thousandth part of the earth's history. The reason for the shortness of the time scale is simple. The uncertainty in the average K-Ar date is about 5 percent. For rocks 1 million years

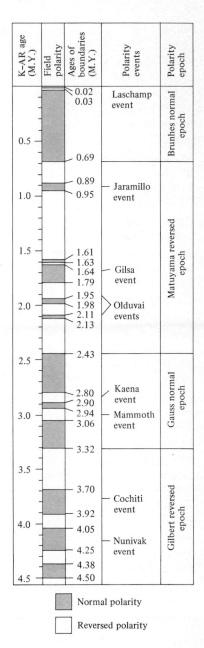

Figure 5-10 The history of reversals of direction of the earth's magnetic field in the past 4 m.y. (*After A. A. Cox.*)

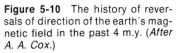

Normal polarity

Reversed polarity

old this represents an error of about 50,000 years. Now the time elapsing between reversals might typically be about 300,000 to 500,000 years. At the 1-million-year age level, then, such reversals could be clearly separated by K-Ar dating. However, a 5 percent error in the age of rocks 5 million years old means an uncertainty of 250,000 years, which is close to the length of the interval one is trying to measure. Thus for times over 5 million years ago, the detailed mapping of the magnetic-field-reversal time scale cannot be done with the aid of K-Ar dating. It is possible, however, with the aid of measurements of magnetic-field variations at sea to extrapolate the time scale shown in Fig. 5-10 backwards to about 70 million years ago, in a way to be explained in Chap. 6. When this is done we see a very similar pattern of field reversals going back 70 million years. We know, however, that this irregular, roughly every-half-million-years reversal of the earth's magnetic field has not been a permanent feature of the earth's history. In a large number of analyses of rocks ranging in age from 230 to 280 million years old, no reversals have been found. Thus it appears that during that period, about 50 million years, the earth's field was much more stable than it is now. It is as if the magnetic field is stable for many tens of millions of years and then bursts into irregular oscillations for at least as long and then becomes stable again. This behaviour has not yet been explained and is a fascinating problem for the theoreticians. Apart from the reversals occurring roughly every half million years or so, little episodes called *magnetic events* (see Fig. 5-10) occur when the earth's field, having been stable awhile, suddenly reverses itself and then almost immediately (that is, within a few thousand years) flips back to its original orientation.

The study of reversals of the earth's field is one of the most intriguing and rewarding in the whole of earth science. Obviously the occurrence of the reversals themslves is fascinating to any scientist. But regardless of why or how they occur, their very existence has enabled us to unravel the evolution of the face of the earth and definitively resolve a 50-year-old controversy. Discussion of this is left to Chap. 6. However, here we now consider the influences field reversals may have had on the course of biological evolution on this planet.

GEOMAGNETIC FIELD REVERSALS AND BIOLOGICAL EVOLUTION

High-energy charged particles, such as protons and mesons, can cause damage to the genes found in living cells. The offspring of irradiated life-forms will accordingly display abnormalities which usually will be a handicap to them in the struggle for survival. We should expect, therefore, that if from time to time the earth were bathed in high fluxes of radiation, these episodes would play a very significant role in the course of the evolution of life on earth. R. J. Uffen in 1963 proposed that the earth is in fact subjected to such genetically crucial showers of radiation every time the earth's magnetic field goes through a reversal of direction. We have just seen that the earth's magnetic field shields our planet from the direct attack of cosmic rays and the solar wind by trapping the various incoming charged particles in the Van Allen radiation belts. But during a reversal the field decays almost to zero strength, and this magnetic shield would thereby disappear for anything from 2,000 to 5,000 years. Uffen consequently argued that during these unprotected millenia very significant genetic changes could be caused by the now undeflected stream of cosmic rays. He proposed that the frequency of field reversal was therefore a very important factor in the course of evolution. This imaginative suggestion aroused much interest for a time, but then it was pointed out by Black and others that the earth's atmosphere itself is a most effective shield against cosmic rays and that the removal of the earth's magnetic field would produce a negligible increase in the incidence of cosmic rays at ground level. Support for Uffen's hypothesis naturally sagged as a result. In recent years, however, new experimental observations have shown that field reversals may in fact be closely connected with evolution but not, evidently, via the mechanism tendered by Uffen. It has been found that cores taken from the sediments at the bottom of the oceans preserved a clear record of the reversals of the earth's magnetic field which have occurred in the past few million years. What has now been observed is that in these same cores is also preserved a record of the levels at which certain species have become extinct. Furthermore there is a strong correlation between the positions in the cores at which extinctions of life occur and those positions where the field reversals are seen. The results found by J. D. Hays of Columbia University, New York, in 1971, make a good illustration. He painstakingly analysed 28 ocean cores and found that preserved in these was the evidence that 8 species of the unicellular microorganism Radiolaria became extinct in the 2.5 million years covered by the cores. Six of these extinctions occurred in the cores at the same level as are found magnetic-field reversals. Now the probability of this correlation happening purely by chance was shown by Hays to be about 1 in 700. Reasonably, therefore, it can be concluded that reversals of the earth's magnetic field may cause the extinction of some species and that the geomagnetic field might accordingly play an important role in the path of evolution. But if the cosmic-ray mechanism of Uffen is not effective, what could the connection be between field reversals and extinctions of life? One possibility discussed by Harrison is that the magnetic-field changes might trigger climatic changes which might cause extinctions. A more intriguing speculation was raised by Hays himself and by I. Crain, who suggested that the important factor may in fact be the direct effect on living organisms of the changes in the earth's magnetic field itself. They pointed out that a number of experiments in the past decade have indicated that several creatures may be distinctly influenced by weak magnetic fields comparable with the earth's present field. Mud snails, flatworms, and fruit flies have all been reported to be sensitive to weak magnetic fields like the earth's.

While it is as yet too early to be sure that these few interesting correlations of field reversals with extinctions of life are significant, the possibility is a real one and is currently being vigorously investigated.

ORIGIN AND MECHANISM OF THE EARTH'S MAGNETIC FIELD

Despite the fact that the earth's magnetic field has been studied intently for many years, there is still no complete theory of its origin. One's first instincts probably would be to speculate that the earth was a great big magnetized ball, just as Gilbert did in *De Magnete*. After all, many of the rocks we find on the earth's surface are magnetized. Indeed, did man not discover magnetism itself in rock? But one's instincts would be wrong. The temperature rises quite quickly within the earth so that by a depth of 20 or 30 km below the surface a temperature is reached at which magnetic materials lose their attractive powers. The earth's deep interior is therefore not a *permanent* magnet. Furthermore, the outer shell, which is cool enough to remain magnetized, is nowhere near strongly magnetized enough to cause the field we measure at the surface.

What is currently considered the most likely solution to the problem was proposed in 1946 by Elsasser. In Chap. 2 we saw that the outer core of the earth is fluid and is probably composed of iron and nickel. Elsasser suggested that a combination of liquid flow and electric currents in this outer core might be able to provide the field that guides our compasses at the surface. The problem, however, is highly complicated mathematically and has so far defeated the efforts of Elsasser, Bullard, and others to find a complete solution. Some flavour of these attempts may be extracted from the following simple model of core motions and electric currents.

Instead of considering the undoubtedly complex motions of the liquid outer core, we substitute a simple circular copper disc. Now it was discovered about 150 years ago by Michael Faraday that when such a conducting disc is spinning about an axle *in an applied magnetic field*, a voltage difference is produced between the axle and the rim of the disc. That is, if we connect one end of a loop of wire to the axle and the other end to the wheel's rim, we find that current flows through the loop, as is shown in Fig. 5-11. Now this current generates its own magnetic field, and if we position the loop correctly with respect to the spinning disc, this new magnetic field can add to the original. If the original applied magnetic field is now allowed to decrease to zero, it is found that, *if we spin the disc quickly enough,* the current keeps flowing through the loop and a steady magnetic field is maintained by this current. We therefore now have a physical system which generates a magnetic field from nothing but a copper disc and some

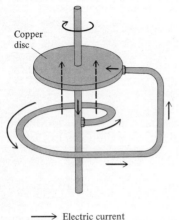

Copper
disc

→ Electric current
--→ Direction of magnetic field   **Figure 5-11** Faraday's disc dynamo.

wire, provided we supply energy continually to spin the disc and
initially have a small applied magnetic field to get things going.
The analogy with the earth is made by saying that the spinning
disc and the current flowing through the loop are equivalent (in a
simple-minded way) to the flow of liquid and electricity in the
outer core. And so if we have a small triggering magnetic field
initially and an ever-present energy source continually driving the
liquid motions, we claim that it seems reasonable that the earth's
core can act as a Faraday disc dynamo and maintain a steady
magnetic field of the type measured at the earth's surface. If we
accept this elementary model, we are now faced with two
problems: Firstly, we must supply the initial triggering magnetic
field; secondly, we have to identify the requisite energy source.
The former is probably not a major problem. The small triggering
magnetic field could no doubt be provided by electric currents
caused by some sort of primitive natural battery action brought
about by variations in chemical composition and temperature in
the earth's deep interior. The source of the energy needed to
drive the essential motions of the liquid outer core is more
difficult to identify, and indeed, agreement has not been reached
on this problem. If these motions are thermally driven convection
currents, we have to produce this heat. One's first appeal in this

case would be to radioactivity, which is present in the earth's surface rocks and certainly generates heat. The snag here, however, is that the important radioactive elements U, Th, and K do not like to congregate in a metal such as the core is presumed to be. They much prefer to be concentrated in silicate rocks, and it seems unlikely that there is enough radioactivity in the core to drive the dynamo. Another possibility, first suggested by Verhoogen, is that the solid inner core has been slowly growing and that the heat released by the steady solidification of an inner core drives the dynamo. Another suggestion is that the slight rocking motion of the earth as it moves around the sun (its *precession*) stirs motions in the outer core just as the contents of a bottle of scotch or Coke may readily be set in motion by gently rocking the bottle.

While it is distressing that the source of the energy needed to drive the dynamo is hard to locate, and although it is disconcerting that rigorous solutions of the mathematics of a more realistic dynamo have not been found, the dynamo theory of the origin of the earth's magnetic field is widely accepted by earth scientists. No serious alternative has yet been proposed. Furthermore, coupled disc dynamos have been shown to be capable of reversing their polarity in a manner which is very reminiscent of the earth's actual field reversals. The dynamo theory, therefore, seems to be here to stay.

Chapter 6

Continental Drift and Plate Tectonics

The idea that the continents have drifted many thousands of miles across the earth's surface is not new. In 1668 P. Placet, a Frenchman, published a memoir entitled *La corruption du grand et du petit monde, où il est montré que devant le déluge, l'Amérique n'était point séparée des autres parties du monde.* In 1858, the year preceding Charles Darwin's epochal treatise *Origin of Species,* Antonio Snider published the two maps shown in Fig. 6-1. He suggested that during Carboniferous times the North and South Atlantic Oceans were closed up. Snider was seeking to explain the great similarity of the fossil plants in the European and North American coal measures. However, these and other similarly isolated flashes of inspiration failed to ignite any general interest in this concept. It was not until the first few years of this century that a more widespread awareness began to dawn that continental drift might very well have occurred.

In this, our final chapter, we will outline briefly some of the contributions of the three great pioneers of continental drift theory: F. B. Taylor, H. B. Baker, and A. Wegener. The ideas of these men met with great opposition for many years, although a few brilliant earth scientists such as A. Holmes and A. L. Du Toit embraced and contributed greatly to the theory. Later, we will see how results from the fledgling field of palaeomagnetism in the 1950s suddenly revived interest in drift and how by 1963 the stage had been set for the advent of a remarkable unifying hypothesis. This concept, of Vine and Matthews, brought together, and made harmonious, evidence from three quite different lines of investigation. By 1966 the supporting evidence was so compelling that a revolution in thinking about the earth's crust had been irresistably launched. An account of these happenings and a sketch of the evolution of the ideas of plate tectonics will bring our story to its close.

F. B. TAYLOR: EARLY CONTRIBUTION

On December 29, 1908, Taylor read to the Geological Society of America a paper entitled "Bearing of the Tertiary Mountain Belt on the Origin of the Earth's Plan." In it he argued that the

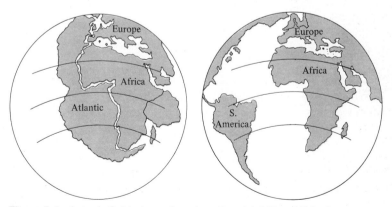

Figure 6-1 Antonio Snider's version of continental drift, published in 1858.

structure and location of the earth's youngest folded mountain system implied large-scale drifting of the continents. North America and Greenland were supposed to have moved away from each other, and Africa and South America were thought to have separated from the Mid-Atlantic Ridge. Taylor wrote several other articles on the subject and also concerned himself with the origin of the forces responsible for drift. He suggested that the moon was captured as a satellite by the earth only about 100 million years ago. The moon was supposed at that time to be much closer to the earth than it is now and the accordingly much greater tidal forces exerted on the earth by the moon were thought to have triggered continental drifting. We mention this imaginative but almost certainly incorrect mechanism because it brings out an important feature of the controversy which surrounded the theories of continental drift for the first half of this century: No convincing mechanism could be found for moving the continents about. All the early investigators were plagued by this problem and much of the most influential criticism was focussed precisely on this point. It is ironic that even now, when continental drift is regarded as an accepted fact, the exact driving mechanism is still in dispute.

H. B. BAKER: SUPERCONTINENT

Baker, in 1911, envisaged that the present continents were at one time parts of a single supercontinent which only 10 or 20 million years ago split up into its present remnants, which drifted quickly to their current positions. His suggested reassembly of the continents is shown in Fig. 6-2. Baker was guided in his reconstruction by the strong complementariness of opposing coast lines and showed how similar rocks and fossils on different continents were in this way brought close together.

A. WEGENER: PANGAEA

Alfred Wegener's initial work, published in 1912, was marginally later than that of Taylor and Baker. His approach was considerably more comprehensive, however, than that of the other early

Figure 6-2 H. B. Baker's postulated continental reconstruction of 1911.

proponents of drift. Furthermore, his presentation and promotion of the ideas involved were so effective that his version of continental drift became a hotly debated topic in the 1920s.

Wegener's ideas were first presented to the public on January 6, 1912, in an address he gave to the Geological Association in Frankfurt am Main entitled "The Geophysical Basis of the Evolution of the Large-Scale Features of the Earth's Crust (Continents and Oceans)." During the same year he published two papers on the subject, but further work then had to be postponed while Wegener participated in a crossing of Greenland in 1912–1913. The First World War broke out in 1914, and Wegener was drafted into a field regiment. Almost immediately he was shot in the arm. No sooner had he returned to active duty than he was struck again, and this time a bullet became lodged in his neck. Whereupon, as Wegener said, "I was able to make use of a prolonged sick leave to furnish a rather more detailed account," and he published in 1915 the first edition of his great work *The Origin of Continents and Oceans*. Revised editions appeared in 1920, 1922, and finally 1929. An English translation of the last is available and still makes exciting reading.

Wegener's reassembly of the continental jigsaw puzzle is shown in Fig. 6-3. He thought that originally there existed one supercontinent, *Pangaea*, which began to fragment in Jurassic times (about 170 m.y. ago), with subsequent dispersal of the continents continuing to the present. The continents were envisaged to be drifting about the earth's surface much as icebergs move across the ocean. Wegener never found a satisfactory driving force for this motion. He was confident, however, that such motions had occurred. He reasoned that if flow could occur in the upper mantle to allow continents to sink and rise during ice loading and unloading (as we saw in Chap. 3), then flow could also occur to allow the continents to move horizontally. In 1930, when Wegener was 50 years old, his exciting life was cut short on the inland icecap on another Greenland expedition.

A. HOLMES: CONVECTION CURRENTS

When a liquid is heated from below, a system of what are called *convection currents* is set up. Fundamentally what happens is that material at depth, which is a little hotter than its surroundings, expands because of the heat and consequently goes down in density. Since this little volume we are considering is now lighter than the surrounding liquid, it will float upwards. In this way ascending currents are set up. Complementary descending currents are established in precisely the reverse fashion, by cooling of the liquid at the surface, with consequent density increases and sinking of now relatively heavy material. The exact pattern of convection cells which is set up in any particular situation is

Figure 6-3 Wegener's original supercontinent Pangaea and its dispersal. *(From "The Origin of Continents and Oceans," Dover, 1966.)*

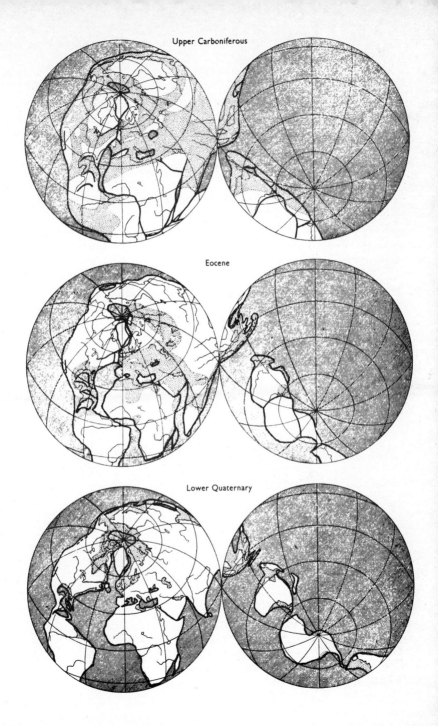

Upper Carboniferous

Eocene

Lower Quaternary

always difficult and frequently impossible to predict, however. A good idea of the complexity of the patterns involved may be easily grasped with the aid of a freshly made cup of black instant coffee. Because of the rapid cooling at the surface, a temperature gradient is set up in the coffee, the liquid at the bottom being hotter than that at the top. Convection therefore sets in. The intricate series of cells within the coffee may now be made visible by very carefully pouring a small amount of milk into the liquid. Before the milk has become completely mixed with the coffee, fine particles of milk in suspension are carried along by the convection currents and an ever-changing pattern of cells can be easily seen in the coffee surface.

Now the earth's mantle is not a cup of coffee. The fact that it transmits seismic shear (S) waves, as we saw in Chap. 2, shows that it is not a simple liquid. However, earthquake waves travel quickly through the mantle, and it is on this time scale that we say the mantle is essentially solid. Continental drifting, in contrast, occurs over tens of millions of years, and on this time scale it seems very reasonable to suppose the mantle might flow, just as pitch is a "solid which can flow."

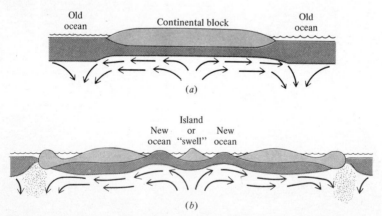

Figure 6-4 Holmes' 1928 version of ocean floor spreading. The continent of (*a*) is split by a convection current which rises and spreads out beneath it. In (*b*) we see how new ocean floor is formed between the drifting fragments while old ocean floor descends into the mantle at the continental edges.

In 1928 A. Holmes pursued the idea that convection in the mantle was responsible for continental drift, and in Fig. 6-4 we reproduce his illustrated mechanism. Holmes speculated that a convection current rising under a continent would spread out laterally, causing the continent to break, and that the diverging lateral currents would then carry the fragments apart. Descending currents occurred at the edges of the continents. Holmes supposed that new ocean floor between the separating fragments was erupted as basalt lava by the upwelling current. We will see shortly that Holmes had come astonishingly close to modern *ocean-floor-spreading* ideas.

A. L. DU TOIT: LAURASIA AND GONDWANALAND

From 1921 onwards Du Toit made numerous useful contributions to drift theory, particularly emphasizing the way it brought the geological histories of South America and Africa into harmony. In one fundamental point he differed from Wegener. Where the latter proposed that one single supercontinent, Pangaea, had existed originally, Du Toit championed the concept that there had been two supercontinents, *Laurasia* and *Gondwana*. The northern mass, Laurasia, comprised North America, Greenland, Europe, and Asia. South of this lay Gondwana, made up of South America, South Africa, Antarctica, Australia, New Zealand, India, Ceylon, and Madagascar. These two continents fragmented, interacted with each other, and eventually yielded the present continental distribution. There is still debate over the relative merits of Pangaea on the one hand and Laurasia and Gondwana on the other, although it should soon be possible to distinguish between them.

CHALLENGED AXIOMS

While we have said that the drift theory met with considerable opposition, we have not explicitly stated what beliefs Wegener and the other drifters were challenging. What they were up against, in fact, was a belief in a far more static earth. It was widely held that the earth had begun its life in a molten state and

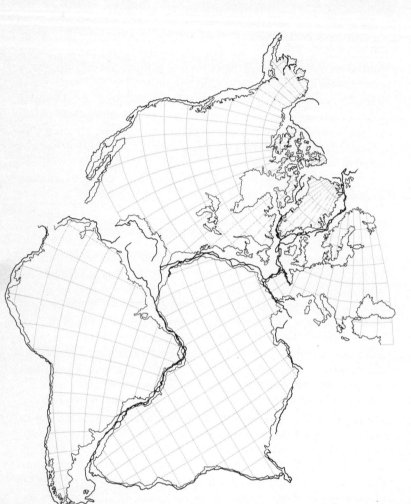

Figure 6-5 Computer-assisted fitting of continents produced by Bullard, Everett, and Smith. The continental boundaries were taken off-shore at the 500-fathom level. A similar fit in the pre-computer era had been found by W. S. Carey. (*Courtesy of Sir Edward Bullard.*)

had been cooling ever since. Since it is a general law that a cooling body shrinks, it was therefore argued that the earth had been shrinking throughout its history. Now the outer part of the earth would have quickly cooled down, essentially as much as it ever would, whereas the interior would naturally cool far more slowly. Consequently the body of the earth would be shrinking beneath the brittle crust. Then, just as the skin of a drying apple becomes all wrinkled and furrowed in adjusting to the shrinking body of the fruit, it was imagined that great folds would appear in the earth's crust as it tried to adapt itself to the dwindling radius of the interior. In this way, it was believed, were the great continental mountain chains built throughout history. The ocean basins were regarded as primeval, permanent features, and while continents could bob up and down under varying ice loads they certainly did not move across the earth's surface. The only serious horizontal motions allowed were those occurring during the mountain-building episodes just mentioned, when a continent could be bent and some crustal rocks could be thrust for a few miles over others.

It is interesting, therefore, at this stage to look a little more carefully at why the early drifters challenged this conventional wisdom. When we do this we find that a fundamental driving inspiration was the extraordinary jigsaw puzzle fit of the continents. While the fit was not perfect, it was considerably better than was admitted by traditional earth scientists. We have already seen the different attempts at reassembly by Baker and Wegener (Figs. 6-2 and 6-3). In Figs. 6-5 and 6-6 we present a picture of how modern electronic computers gave their solutions to the jigsaw puzzle. In 1965 Bullard, Everett, and Smith essentially fed first the northern continents and then South America and Africa to the Cambridge computer and told it to piece together the best-fitting results. By this we mean they instructed the computer to assemble the pieces in the way which produced the least amount of overlapping and underlapping at the points of contact. How extremely well this could be done is illustrated by Fig. 6-5. Later, Smith and Hallam played the same game with another computer using South America, Africa, Antarctica, Australia, India, and Madagascar and came up with the result in Fig. 6-6. While a reasonable fit was found, it was less convincing than the earlier fittings because the southern continents have smoother outlines and the pieces do not lock back together again in so convincingly unique a fashion. However, both these modern

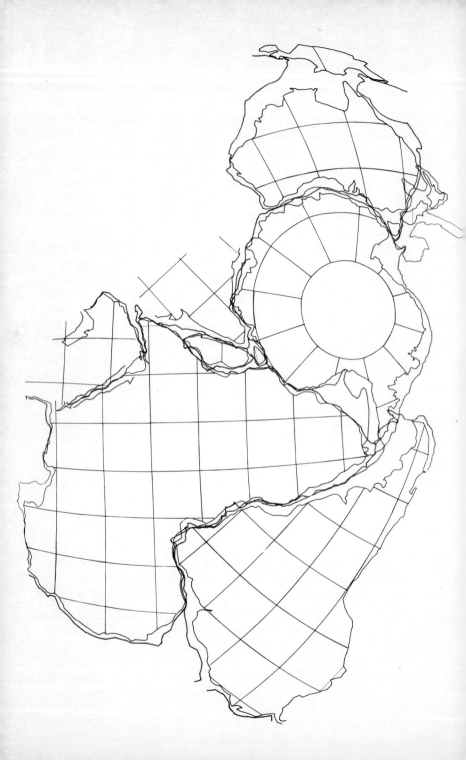

fitting exercises, and others like them, confirmed that on geometrical grounds alone the continents could well have once been part of one or two supercontinents. They also confirmed Du Toit's suggestion that the edges of the continents to be fitted should not be drawn at the coastlines. Rather the continental margins should be taken offshore at, say, the 500-fathom level, which would be well down the continental slope we illustrate in Fig. 6-24 and which would be a better measure of the boundary of a continent than is the coastline. After all, the coastline is a very impermanent thing, depending as it does on the balance between water in the oceans and water locked up as ice in the polar regions and Greenland.

A second type of argument used by the drift pioneers was that if one assumed continental drift had occurred, then certain interesting features in the distributions of fossils were readily explained. For instance, sedimentary rocks of Carboniferous, Permian, and Triassic ages (see Chap. 4) on continents now separated by a large ocean might have virtually undistinguishable assemblages of fossils. But then one might find that succeeding sediments of the Jurassic, Cretaceous, and Tertiary periods had fossil families which showed divergencies between the two continents which became more marked the younger the rocks. Paleontologists of the late nineteenth century had sought to resolve such effects with the aid of *land bridges*. They would have explained the situation we have just described by proposing that, while the two continents are now separated by an ocean, in Carboniferous, Permian, and Triassic times a vast tract of land joined these two continents. One thus had a kind of supercontinent with free interchange and migration of land life. Then, in the Jurassic age the central area began to sink, allowing the sea to encroach and begin the isolation of the two present-day continents. As time went on, the sinking land "disappeared" and life on each isolated continent naturally followed its own path; thus we would have an explanation of the present observations. Wegener, however, who was very interested in these fossil distributions, said such a mechanism was, in general, completely unacceptable. He appealed to the concept of isostasy which we

Figure 6-6 Southern continents fitted together by Smith and Hallam. (*Courtesy of A. Gilbert Smith.*)

discussed in some detail in Chap. 3. There we saw that the earth is essentially in hydrostatic equilibrium. In particular the continents, having large amounts of comparatively light granitic rocks, are floating serenely like logs in water and have no inclination to sink without trace into the earth's interior, although we know they can be depressed somewhat by great ice loads. It seemed, therefore, to Wegener that these hypothetical, convenient, massive land bridges could not be gotten rid of so easily as the paleontologists desired. Despite the cogency of this argument, the paleontologists did not, in general, follow Wegener. Instead they said that perhaps the land bridges had really been very narrow strips linking continents and that any required homogenizations of life had been achieved via these few but critical routes. Thus about 2 or 3 million years ago a sharing of mammal life began between North and South America when a land bridge, the current Isthmus of Panama, was established. Prior to this, quite distinct families of mammals had developed on the unconnected continents. Such *demonstrable* land bridges were not to be denied, but Wegener held to his main thesis that the wholesale disappearance of such bridges throughout history was highly unlikely. Rather than using land bridges, he said, why not merely consider that the two continents which are now isolated were, at the time of the postulated land bridge, actually in contact.

A well-known fossil-distribution effect is illustrated in Fig. 6-7. Fossils of the reptile Mesosaurus have been found only in South Africa and Brazil. This little creature, perhaps 18 in. from tip to tail, was swimming in these areas about 270 million years ago. If Wegener's continental reconstruction is adopted, then South America and South Africa were one big continent when Mesosaurus was alive, and his hunting grounds, which are now two widely separated areas, formed one single territory. It was more than 100 million years after the extinction of Mesosaurus that its old haunts began to split roughly down the middle and drift apart. Those who were skeptical of drift, on the other hand, argued that perhaps Mesosaurus was able to swim across the South Atlantic. The snag with this argument is that if Mesosaurus was such a fine swimmer, then it should have spread much more widely along the African and South American coasts than it did.

THE FALLOW YEARS

Following the excitement and controversy of the 1920s, little progress was made in the 1930s and 1940s towards convincing a majority of earth scientists that continental drift had occurred. This was largely because another theory, contraction of the earth, was already firmly ensconced and because the evidence for the new theory, while plausible, was in reality not definitive. Arguments based on the shape of continents and vagaries in the distribution of fossils were *qualitative* and failed to impress most leading earth scientists, geophysicists in particular. For over 20 years, then, the theory of continental drift made no headway. Very little new supporting evidence was adduced, most of its adherents having to content themselves with rehashing the well-known arguments. A shift in the centre of gravity of opinion could be brought about only with evidence from new fields.

THE DAWNING

The first such new clues came from the renaissance of interest in rock magnetism in the 1950s. As we saw in Chap. 5, a few outstanding individuals had examined the magnetization of rocks in the first half of this century. Some of these had even suggested that the earth's magnetic field had reversed its direction from time to time. Their work, however, had little influence on the general

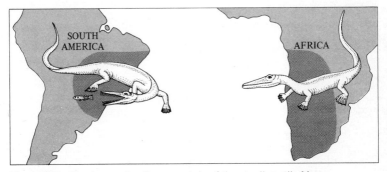

Figure 6-7 The former hunting grounds of the small reptile Mesosaurus are now separated by several thousand miles of Atlantic Ocean. If the continents are fitted back together as in Fig. 6-5, Mesosaurus's haunts become a single territory. (*After A. Hallam.*)

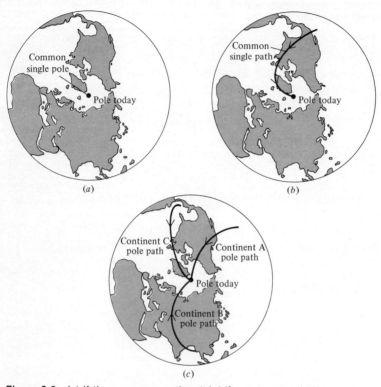

Figure 6-8 (*a*) If there was no continental drift nor any wandering
of the magnetic poles, lavas of various ages from different continents would all have frozen-in compass needles pointing to the
same single North Magnetic Pole. (*b*) If the continents remained
fixed with respect to each other but the magnetic pole wandered,
then rocks of differing ages from one continent would record this
polar-wandering path. Rocks from all other continents would also
record this same single-pole path. (*c*) If, in addition to magnetic
polar wandering, the continents drifted in latitude with respect to
each other, the single-pole path of (b) would be split into one pole
path per drifting continent. The position of the *magnetic pole* today
is shown at the *geographic pole* for reasons given on page 103.

course of earth scientific thinking, and it was not until the second
half of this century that rock magnetism escaped the doldrums
and showed the first glimpses of the power it was to become.

We recall from Chap. 5 that when basalt lava cools, following
its eruption from a volcano, it behaves like a collection of billions

of compass needles which point to the position of the North Pole at the time the lava solidifies. This direction is frozen into the rock and recorded so that, with luck and skill, this position of the pole at eruption may be read today, perhaps hundreds of millions of years after the extinction of the volcano. The record of the position of the North Magnetic Pole (and of course, therefore, also the South) throughout the earth's history is consequently lying filed for inspection in the piles of lavas scattered about the earth's surface. Each lava flow is like a page in some gigantic rock-magnetic diary kept by the earth. On each page the earth noted faithfully where its North Magnetic Pole was that day. Of course we have only the fragmentary remains of the diary, but enough pages had been recovered and translated to build up the following scenario in the 1950s.

Suppose that for the last 600 million years the North and South Magnetic Poles of the earth, on average, remained fixed with respect to the surface of the earth (small deviations would not be important). Imagine further that the crust was fixed relative to the earth's interior; i.e., there was no drifting of continents separately or of the crust as a whole rigid body. Then if we read the position of the North Magnetic Pole as recorded in the basalts of various ages erupted on different continents, *we would find*, within the limits of experimental error, *the same value for the North Pole position from each lava*, regardless of its age (see Fig. 6-8a) and its continent of origin.

Now consider what we would find in our magnetic diary if, as before, there was no continental drift, but now the magnetic pole had moved slowly large distances with respect to the crust. Then we would find from the pole positions recorded in the lavas of North America a clear trace of the path of the pole. The lavas from all other continents would also record this identical path. This is illustrated in Fig. 6-8b.

Suppose we now modify the last situation by allowing that the continents drifted to new latitudes while the pole was also wandering. In this case our previous rock-magnetic record would be split, from one single polar-wandering path which was deduced from the records of all continents, into a different polar-wandering path for each continent, as we show in Fig. 6-8c.

When this reading of the magnetic diary was first done in some detail, by the Newcastle group of Runcorn, Irving, Creer, and Collinson, they immediately found separate pole paths for distinct continents and concluded that there must therefore have been continental drift. Some actual polar-wandering paths are shown in Fig. 6-9. The displacements of the pole paths were roughly consistent with the most prominent features of the drift theory, such as the opening of the Atlantic in the past 200 million years, and it may be said that these measurements in the 1950s heralded the dawning of a new understanding of the earth's surface behaviour.

OCEAN-FLOOR SPREADING

Continental drift now became an acceptable topic of conversation again, and in 1960 an important paper on the subject was written by H. Hess, an eminent Princeton geologist. In this, Hess was largely restating Holmes' model that we illustrated in Fig. 6-4, but the concept was presented in the light of the greater knowledge of the ocean floors which had been gained by 1960. Thus by this date it had been established by seismic work at sea that the earth's crust beneath the oceans is completely different from the continental crust. It is chemically different, being largely basaltic in composition, and it is a factor of 4 or 5 times thinner on average. Furthermore it had been found by Heezen and others that the enormous submarine mountain chain in the middle of the

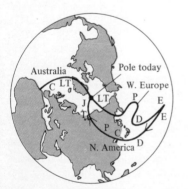

E – Cambrian pole position
D – Devonian pole position
C – Carboniferous pole position
P – Permian pole position
T – Triassic pole position
J – Jurassic pole position
LT – Lower Tertiary pole position

Figure 6-9 Polar-wandering paths recorded by rocks from North America, Western Europe, and Australia. (*After E. R. Deutsch.*)

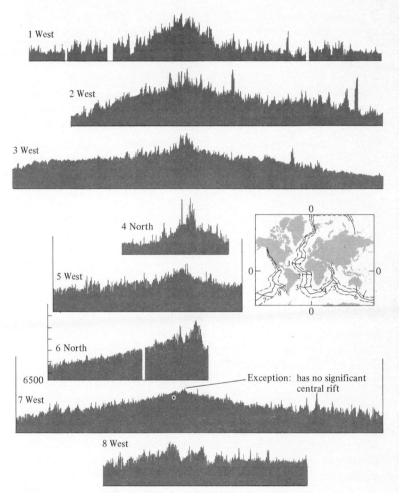

Figure 6-10 Usually a deep rift runs down the middle of the mid-ocean ridge. The locations of the profiles are shown on the small map of the world. The base-lines of all the profiles are 6,500m below sea-level. (*After B. C. Heezen.*)

Atlantic known as the *Mid-Atlantic Ridge* is merely one portion of a globe-girdling feature. These mid-ocean ridges also usually display a deep cleft or rift at their centres (see Fig. 6-10) and are the locus of most of the earth's shallow earthquake foci. Deep troughs had also been carefully mapped in the ocean floors at places along the rim of the Pacific Ocean where the ocean floor seemed to bend down, producing great depths of ocean (Fig. 6-11). In addition, no rocks over about 200 million years in age had been recovered from genuine ocean floor. Hess took all these factors into account and proposed that continental drift had occurred via an *ocean-floor-spreading mechanism*. He envisaged that the mid-ocean-ridge system represented the locus of upwelling of convection currents in the mantle. The ocean floors themselves were imagined to be participating in the convective circulation. Thus they were supposed to be moving symmetrically away from the mid-ocean ridges like gigantic conveyor belts. Naturally, if the ocean floors were moving away from the central ridges, *new ocean floor* had to be created to fill the gap left behind. We thus have the picture of basaltic ocean floor material being erupted along the mid-ocean-ridge system, cooling down and solidifying, and then moving off to both sides as new ocean floor. This conveyor-belt process of ocean-floor spreading is illustrated in Fig. 6-12. The continents, then, on this model, have drifted about the earth's surface by being passively carried along by the conveyor system. That is, they did not have to plough *through* the rocks of the ocean floor as Wegener wanted. Since the individual continents have drifted apart by the order of 1,000 km or more in 100 to 200 million years, we see that Hess' conveyor belts must have been moving at speeds of the order of 1 to 2 cm/year, if the motions were reasonably steady. There is an immediate problem now, however, of what to do with all this ocean floor which is steadily being produced at the mid-ocean ridges and ferried away to the sides. There is no evidence of its accumulating anywhere on the earth's surface. Two reasonable possibilities might be considered. We might suppose, as B. Heezen did in a paper published a little before Hess', that the earth was slowly expanding. In this case the earth's surface area would of course be steadily increasing, and as Heezen suggested, the creation of new ocean floor might be able to keep up with this and in a sense

Figure 6-11 The Pacific Ocean is rimmed in many parts by stretches of extraordinarily deep water. These are shown as heavy black lines. (*After D. L. Turcotte and E. R. Oxburgh.*)

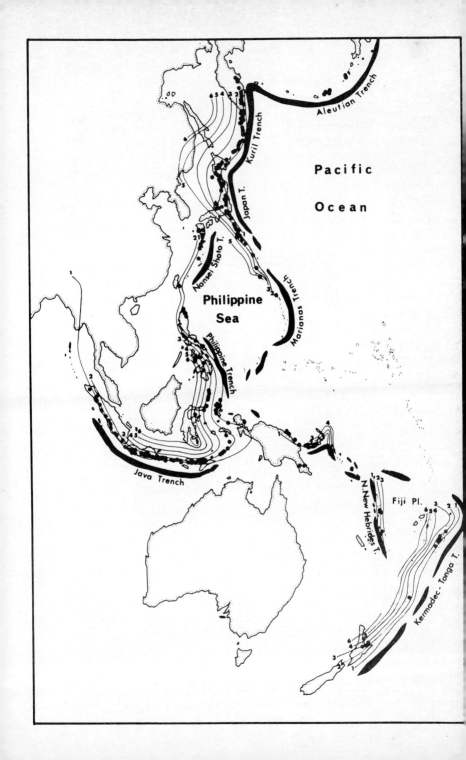

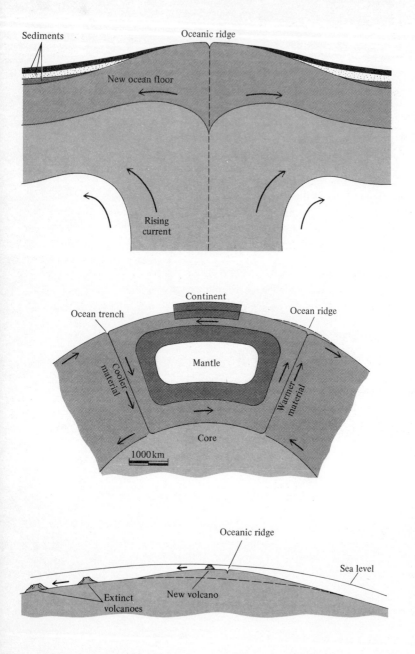

provide new paving for the newly appearing surface area. This would represent a kind of Parkinson's law, in which ocean floor expands to fill the space available. Alternatively, as Hess suggested, we might get rid of "unwanted" ocean floor by allowing it to sink back into the earth's interior at sites marked by the great ocean trenches. Of these two explanations, Heezen's has been rejected because it requires the earth to have undergone a much greater expansion than seems justifiable on other grounds. Hess' concept, on the other hand, of resorption of ocean floor in the trenches has now come to be fully accepted. We thus have the overall concept of ocean floors being created at ridges, gliding across the earth, and being returned to the depths at trenches. Ocean floors are thus continually being created and destroyed, and because of the earth's dimensions and the speed of the conveyor belts, we cannot expect to see any significant amount of ocean floor older than about 200 million years. If you like, "the slate of the ocean floor is wiped clean" every 200 million years. In complete contrast the continents, as we saw in Chap. 4, record at least 3.7 billion years of history. Presumably the low density of continental rocks keeps them afloat and prevents their suffering the fate which awaits the drifting ocean floors.

This theory of Hess was to be one of three quite different threads which were woven together in 1963 by F. Vine and D. H. Matthews. To prepare ourselves for the Vine-Matthews synthesis, we will now quickly look at the two remaining strands which came from the fields of oceanography, geomagnetism, and geochronology, which were all in a state of great expansion in the early 1960s.

THE TWO MAGNETIC THREADS

In the late 1950s many measurements were made at sea of the strength of the earth's magnetic field. Magnetometers of great

Figure 6-12 Hess' model of ocean-floor spreading. New ocean floor erupted at the mid-ocean ridge is ferried away to the sides. Hess also indicates how a chain of volcanic islands could be created as newly formed peaks are carried away from a source on the ridge.

precision were used which could detect variations in field strength of about 1 part in 100,000. When the data from these expeditions were plotted it was found that there are linear patterns of highs and lows in the earth's field strength. These corrugations in the earth's field are shown in Fig. 6-13. While they represent changes in the strength of the field of about merely 1 percent as one goes from a high to a low, they are easily measureable when we have the precision mentioned earlier. For 4 or 5 years they were well-known features, but they defied correct explanation.

The second important magnetic factor has already been described in detail in Chap. 5. It was the verification that the earth's magnetic field does reverse its direction and the establishment of an approximate time scale of these reversals by combined paleomagnetic and K-Ar dating studies on basalts.

THE VINE-MATTHEWS HYPOTHESIS

While the polar-wandering results of the 1950s were awakening earth scientists to the fact that continental drift should be given a serious reexamination, and the appearance of Hess' paper was undoubtedly a reflexion of this, it is still true to say that by as late as 1963 most earth scientists had not appreciated that a revolutionary idea was about to be ushered quietly in. We say quietly because the full significance of the Vine-Matthews hypothesis did not become widely understood until 1965.

In 1963, F. Vine and D. H. Matthews of Cambridge University had a brief article published in *Nature*, an English scientific journal devoted largely to the rapid publication of new ideas. In this paper, these two young Englishmen took Hess' ocean-floor-spreading concept and showed how, with the aid of the idea of reversals of the earth's magnetic field, they had a natural explanation of the great belts of linear magnetic anomalies found at sea. Like so many great ideas, theirs was beautifully simple.

We use Fig. 6-14 to illustrate the argument. Imagine a roughly straight stretch of mid-ocean ridge, say 600 km long. The ocean floor to either side of this ridge is retreating from the ridge at about 2 cm/year. Now the new ocean floor which is being created along the ridge centre is basaltic rock which has cooled from the molten state in the earth's magnetic field. It therefore

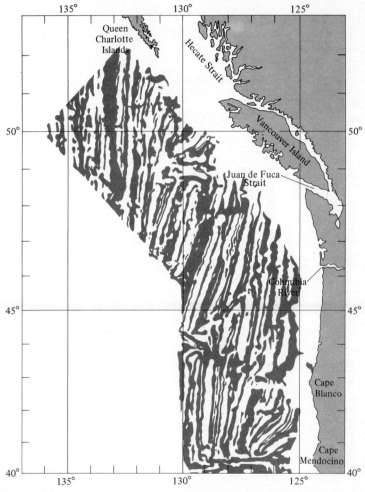

Figure 6-13 Linear magnetic anomalies at sea. The earth's magnetic field is higher than average in the dark areas, below average in the light areas. (*After A. D. Raff and R. C. Mason.*)

has become magnetized in the direction of the earth's field, as we discussed in Chap. 5. As this magnetized ocean floor recedes symmetrically on both sides of the ridge, we have, in essence, paved both sides of the ridge with a strip of magnetized rock which parallels the length of the ridge. We now suppose that the

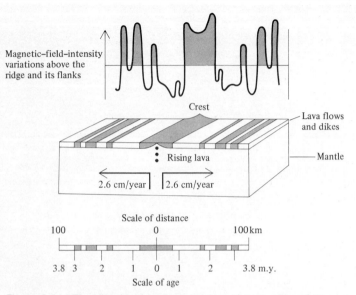

Magnetic-field-intensity variations above the ridge and its flanks

Crest

Lava flows and dikes

Rising lava

Mantle

2.6 cm/year | 2.6 cm/year

Scale of distance

100   0   100 km

3.8 3   2   1   0   1   2   3.8 m.y.

Scale of age

Figure 6-14 The Vine-Matthews mechanism of generation of the linear magnetic anomalies. (*After J. Tuzo Wilson.*)

earth's magnetic field reverses in direction and the previous spreading process is repeated. Everything follows exactly as before, except that now the new strips on either side of the rift are oppositely magnetized to the earlier produced strips, which are, of course, now farther from the ridge centre. This process is now imagined to be repeated for millions of years, new strips-ofocean floor being steadily produced, having first one direction of magnetization and then the opposite. The final result is that the ocean floor is covered with strips of *normally* and *reversely* magnetized rock, the stripping being parallel to the ridge and symmetrical with respect to it. Here then is the explanation of the linear magnetic anomalies in the earth's magnetic-field strength found at sea. For if we were to cruise in a ship at right angles to the ridge, and hence at right angles to the magnetic pattern on the ocean floor, we would find a slight increase in the earth's average field if we were above a normally magnetized strip of ocean floor. Whereas if we were above a reversely magnetized strip, the magnetism in the ocean floor would produce a small field which opposed the earth's present field, and hence we would find a low in the field. And so, on any such nautical traverse at right angles to the ridge, we would find a succession of highs and lows in the earth's total field strength, and by doing many such traverses we would map out the well-known linear anomalies.

At the time of Vine and Matthews' original paper in 1963, the symmetry of the pattern about the ridges had not been documented. The field-reversal time scale (see Chap. 5) was also in a somewhat primitive state. However, the scale of the pattern of anomalies in the magnetic field at sea was what you would expect if the ocean floor conveyor belts moved at 1 to 5 cm/year and if the field reversed about every 300,000 years. There was also a well-known high in the field above the oceanic ridges, which is expected on this theory. Within 2 years, however, the most important features of the magnetic-reversal time scale had been clarified and many more magnetic profiles at sea had been taken. Thus Vine and J. Tuzo Wilson of Toronto were able to make a convincing 1-to-1 comparison of features of the belt of anomalies near the Juan de Fuca Ridge just off the West coast of North America with the pattern of reversals of the geomagnetic field. J. R. Heirtzler and his Lamont colleagues meanwhile found that the large belt of anomalies south of Iceland lay *symmetrically* about the Reykjanes Ridge, again in conformity with the new hypothesis (Fig. 6-15). It was now clear, in fact, to the oceanographers that the Vine-Matthews hypothesis was basically correct. Furthermore, in an amazingly short span of time they had published magnetic profiles for portions of all the major oceans. These revealed the same striped wallpaper pattern of highs and lows hung symmetrically about the various ocean ridges of the world. The only serious differences between corresponding portions of the patterns from different oceans were differences of scale. This meant that some oceans were definitely spreading away from their formative ridges much more quickly than others. The anomaly patterns in the fast-spreading oceans would obviously have broader stripes. This is illustrated in Fig. 6-16. Finally, by comparing these oceanic striped-wallpaper patterns with the striped pattern of Fig. 5-10, which portrays the sequence of reversals of the past 3.5 million years, it was possible to say that such and such a black stripe on the oceanic pattern obviously corresponded to a particular one on the reversal time scale. But if the centre of this oceanic black stripe were located, say, 20 km

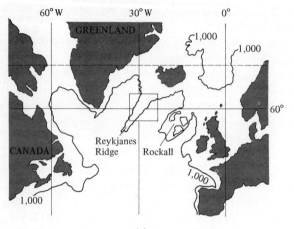

(a)

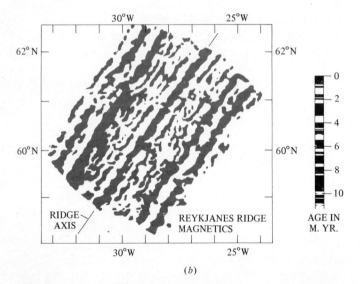

(b)

Figure 6-15 J. R. Heirtzler and his colleagues found that the magnetic anomaly pattern along the Reykjanes Ridge, south of Iceland, matched the magnetic-field-reversal time scale. Just as importantly they noted that the anomaly pattern was *symmetrical* about the ridge. The 1000-fathom depth contour is shown in part (a). (*After F. J. Vine.*)

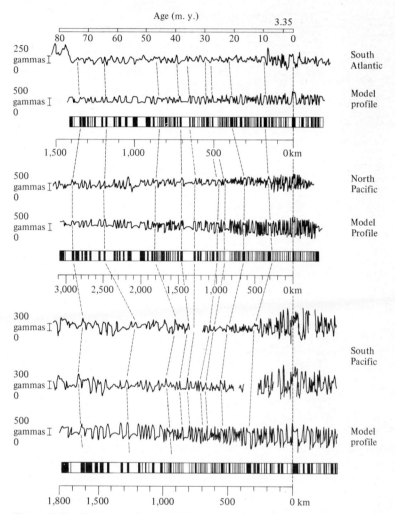

Figure 6-16 The magnetic anomaly patterns for various oceans are very similar except for the scales. The pattern is stretched out over fast-spreading oceans such as the North Pacific and more closely spaced over a more sedately growing ocean like the South Atlantic. The model profiles shown were generated from the reversal time scales beneath them and compare very well with the actually observed magnetic profiles shown above them. The North and South Pacific show some variation in spreading rates with time. The South Atlantic appears to have spread at an approximately constant rate. (*After J. R. Heirtzler, F. J. Vine, and others.*)

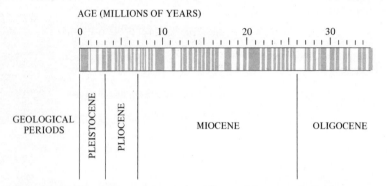

Figure 6-17 The history of reversals of the earth's magnetic field during the past 70 m.y. as deduced from magnetic anomalies in the South Atlantic, assuming a constant ocean-floor-spreading rate. (*After J. R. Heirtzler and others.*)

from its parental ridge, while the centre of the corresponding stripe on the reversal time scale corresponded to a state of the earth's field, say, 2 million years ago, we would be able to say that that particular strip of ocean floor must have drifted 20 km away from the ridge in the 2 million years which elapsed following its production and magnetization. We would therefore have discovered that this part of the ocean floor was spreading at 20 km per 2 million years, i.e., at 1 cm/year, away from its ridge.

Heirtzler and his colleagues carried out this pattern matching for the various oceans and came up with the following approximate spreading rates: Atlantic and Indian Oceans 1 to 2 cm/year; Pacific Ocean approximately 5 cm/year. And now a bonus was presented to these earth scientists. We saw in Chap. 5 that the field-reversal time scale was established (by paleomagnetic and K-Ar measurements) from only the present time back to 3.5 million years ago because of the lack of resolution of the K-Ar technique for rocks much older than this. It was possible therefore to match the pattern of Fig. 5-10 with only the first small fraction of each of the oceanic wallpaper patterns which, in addition, recorded reversals occurring long before 3.5 million years ago. But if we now suppose that this oceanic wallpaper pattern "came off the printer" at a fixed rate in any given ocean, we can say that, for example, if the last 35 km were printed in the last 3.5 million years, then a reversal indicated by the edge of a

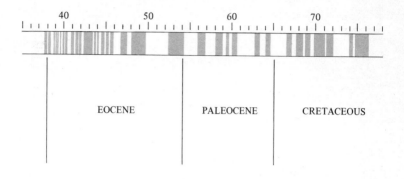

stripe in the oceanic pattern lying 70 km from the ridge must have occurred 7 million years ago. Similarly a reversal recorded in the anomaly pattern 140 km from the ridge must have taken place 14 million years ago; and so on. The field-reversal time scale was thus extended back from the independently established 3.5-million-year-ago limit to over 70 million years ago (Fig. 6-17). The assumption of constancy of spreading rate, or of wallpaper printing, could then be tested by repeating the exercise for the various oceans and their anomalies. When this was done, the same reversal time scale was found, provided some variations of spreading rates with time for the North and South Pacific were allowed. It thus became possible to work out the age of an isolated segment of ocean floor by finding the pattern of magnetic anomalies above it and matching this against the newly expanded field-reversal time scale. In Fig. 6-18 we show the progress that has been made in dating the ocean floor via these anomalies.

TRANSFORM FAULTS

The consistency of the results we have just discussed was so convincing that by the end of 1966 the revolution in thought had been successfully achieved. Continental drift via ocean-floor spreading had become the conventional wisdom. In addition to these critical magnetic results, there was one more influential

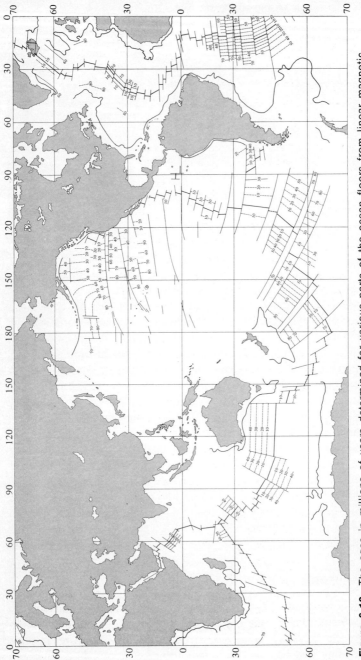

Figure 6-18 The ages in millions of years determined for various parts of the ocean floors from linear magnetic anomalies. The heavy black lines are mid-ocean ridges offset in many places by transform faults. *(After J. R. Heirtzler and others.)*

contribution made, in 1965. This came simultaneously and separately from two Canadians, J. Tuzo Wilson and A. H. Coode.
These two were addressing themselves to the fact that the
mid-ocean ridges are not smoothly continuous features but in fact
are chopped up into relatively short segments offset by many
faults. One short stretch of such a ridge is drawn out separately in
Fig. 6-19 so that we can see more clearly what proved to be so
fascinating about these structures. In the diagram we have two
segments of mid-ocean ridge offset from each other. The line
along which the offset occurs is called a *transform fault.* In
reality, of course, the line is just the surface trace of a vertical
plane. Now let us fix our attention initially on Fig. 6-19*a* and
suppose we are told that a fault plane, such as is marked by *AB*, is
commonly the site of numerous small earthquakes. Then one's
first reaction would be to conclude that the rock to the north of
the fault line has clearly been moved to the west relative to the
rock south of the fault trace in order to displace the two segments
of ridge. So if there are now occasional earthquakes along the
fault, this is merely because the process is being continued and
the earthquakes result from the rather jerky movements as the
ridge segments are offset even further. However, Wilson and
Coode said, this is not so, if you adhere to the theory of
ocean-floor spreading. This is justified in Fig. 6-19*b*, where arrows
have been drawn to show how the ocean floor is moving away
from the two ridges. If, as seems reasonable, spreading occurs
from both segments at the same speed, then it is clear from the
diagram that relative motion between the material north and
south of the fault *AB* exists along only the portion *CD* of the fault
line. This contrasts with our "first reaction" explanation in which
relative motion occurs along the whole of the fault line *AB*. A
difference such as this can immediately be put to observational
test by examining the distribution of earthquake foci along the
fault trace. Such a check shows that the earthquakes occur along
only that portion of the fault lying between the offset ridge
elements; that is, they are confined to *CD* in keeping with the
deduction from ocean-floor-spreading theory. But this is not all.
Ocean-floor-spreading theory predicts another geometrical feature of the earthquakes along the fault, which differs from the
prediction one would make using a prespreading approach. As we

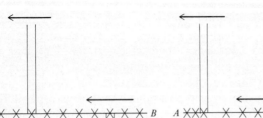

Initially   Later

(a)

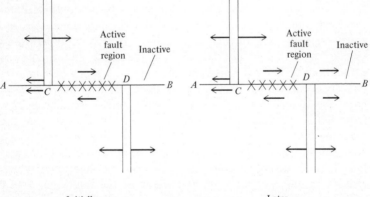

Initially   Later

(b)

Figure 6-19 (a) When ocean-floor spreading from the ridge segments is not allowed, it is logical to suppose that any new motion
along the fault AB will cause the offset between the ridge segments
to increase. Rocks to the north of the fault would be expected to
move to the west, oppositely from those south of the fault. (b) When
ocean-floor spreading from the two ridge elements is allowed, the
directions of motions are seen to be opposite from those predicted
in (a), the current faulting is confined between the offset ridge
elements and, therefore, so are the earthquakes. Earthquake locations in both (a) and (b) are marked by crosses.

have seen, the latter would predict that the motion causing an earthquake along the transform fault would be a sudden *motion of the rock north of the fault plane to the west with respect to the rocks to the south.* In complete contrast, as we see at once from Fig. 6-19*b*, when ocean-floor spreading is assumed to occur from the two ridge elements, not only are the earthquakes predicted to lie only along *CD* but the relative motion at the source of the earthquake on opposite sides of the fault trace is also precisely the reverse of what we have just seen would be forecast on a theory not allowing spreading of the ocean floors. That is, the sudden movement causing an earthquake would be the result of ocean-floor spreading and would entail motion along the transform fault in which *the rock north of the fault moved to the east with respect to the rocks to the south.* The first test of this striking prediction of ocean-floor-spreading theory was performed by L. R. Sykes of Lamont Geological Observatory in 1967. Sykes examined the first motions on the earthquake records taken at observatories sited at various points around the compass with respect to the earthquake foci. This enabled him to calculate the directions of relative motions along the transform faults which had occurred during earthquakes. Without exception, for 10 earthquakes along 2 mid-ocean ridges, Sykes found that the directions of relative motions agreed with those predicted for the transform fault on the ocean-floor-spreading theory. This striking confirmation of a purely theoretical prediction combined with the evidence from oceanic magnetic anomalies studies to convince almost all earth scientists that ocean-floor spreading has indeed occurred for at least the past 200 million years. Doubtless it will continue for many more hundreds of millions of years. For instance, Africa appears to be breaking up along the great rift valleys. Few would also doubt that ocean-floor spreading went on long before 200 million years ago, but this is much more difficult to show, of course, since the ocean floors along with their magnetically recorded evidence of spreading usually exist for no more than about 200 million years before being returned to the mantle. Such ancient spreading episodes must apparently be elucidated by examining the continents and the evidence they retain of such early ocean-spreading histories.

Wegener, "thou shouldst be living at this hour."

JOIDES

By 1968, it was clear that many features of the earth's past and present behaviour could be beautifully explained in the context of the theory of ocean-floor spreading. However, one obvious but critically important test remained. If the theory were correct, the basalt of the ocean floors should be of essentially zero age near spreading ridges and should be found to be steadily older the further it was from its parental ridge. The testing of this crucial point was carried out as part of the JOIDES project. JOIDES stands for *Joint Oceanographic Institutions for Deep Earth Sampling*, which refers to a project whose development began in 1964 when five American institutions interested in oceanography planned a major series of cruises to obtain samples of the ocean

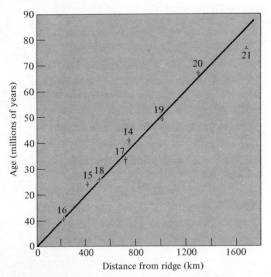

Figure 6-20 The age of the oldest sediment recovered at a drill site is plotted against the distance of the site from the parental ocean ridge. The sizes of the crosses are a measure of the uncertainties in age and distance. Cores 17 and 18 were drilled to the east of the Mid-Atlantic Ridge, the remainder to the west. (*After A. E. Maxwell and others.*)

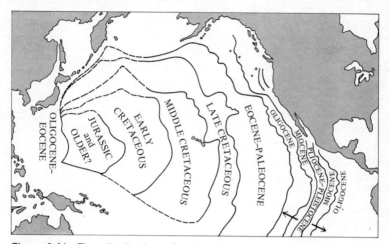

Figure 6-21 The distribution of ages of the oldest sediments found by drilling on leg 6 of the *Glomar Challenger* cruise. The results show a steady increase in age of the ocean floor as one moves away from the East Pacific Rise. Ages in millions of years corresponding to Eocene, Cretaceous, etc., are given in Table 4-3. (*After A. G. Fischer and others.*)

floors by drilling at sea. The research vessel *Glomar Challenger* was outfitted to be capable of drilling through about a 1-km thickness of sediment in water depths up to 5 km. The intention was to collect long cores of the sedimentary cover on the ocean bottom and, if possible, to recover some of the basalt below. These cruises, begun in 1968, have in fact been remarkably successful and have verified the predictions of the theory of ocean-floor spreading. We illustrate the results for the South Atlantic and the Pacific Ocean in Figs. 6-20 and 6-21 respectively. In the first figure, we see a plot of the age of the ocean floor at a drilling site versus the distance of that site from the parental ocean ridge. The points thus produced clearly show the South Atlantic ocean floor is older the further one goes from the ridge. Furthermore, the points lie quite closely about a straight line, which strongly indicates that the ocean-floor spreading has occurred at a steady rate for the past 70 m.y. The slope of this line with respect to the *distance* axis shows that the South Atlantic ocean floor was ferried away from the ridge at about 2 cm/year, which agrees with what had been predicted beforehand from the magnetic anomaly patterns in the area. Furthermore, when the line drawn through the sample points is extended, it goes through the zero-distance mark at zero age, again as predicted. These results, therefore, published in 1970 by Maxwell and his colleagues, spectacularly confirmed the ocean-floor-spreading concepts. The aging of the Pacific Ocean floor as one moves away from the parental East Pacific Rise is shown in Fig. 6-21.

It is worth emphasizing that in these studies the age of interest being determined is that of the basalt layer of the ocean floor at the drill site; i.e., one is not really after the age of the various layers of sediments which might frequently cover this basalt. It is the basalt which was formed fresh at the ocean ridge and was gradually carried away, whereas sediment is continually being deposited at all points of the ocean floors. However, the basalt samples recovered are themselves not directly dated with ease. Normally they would be dated by the K-Ar method. Submarine basalts, however, often contain excessive amounts of argon-40, and so the radiometric approach is difficult. Since the basalt, of course, because of its origin from the molten state, contains no fossils, it also cannot be dated paleontologically. The age of the basalt must therefore be found with the aid of the sedimentary cover. Clearly the layer of sediment *immediately on top of the basalt* at a site must have been deposited after the basalt crystallized, and so if we could date the time of deposition of this particular sedimentary layer, we would have a *minimum* age for the basalt. The time of deposition was in practice determined by identifying the fossils in the sediments and looking up in the geological time scale the time when such creatures were alive. This paleontological approach was necessary because it is extremely difficult to date the time of formation of a sedimenatry layer with the radiometric techniques of Chap. 4. It is this *minimum* age of the basalt, then, which is shown in Figs. 6-20 and 6-21. Some errors will have been introduced by this practice, but the consistency of the results shows that these errors are not likely to be serious and that a sedimentary cover is quickly laid down on newly formed basalt.

PLATE TECTONICS

It might by now seem to the reader that we have reached the end of our story, with the vindication of many of the ideas, and certainly the spirit, of Taylor, Baker, Wegener, Holmes, and Du Toit. The remainder would seem to be largely a filling in of details, a reconciling of the observations from all the various fields of earth science with the continental drift and ocean-floor-spreading theories. Yet this would be incorrect. Another important concept remained to be found. In a sense it was inherent in the ocean-floor-spreading theory we have just described, but the concept is sufficiently important that its specific enunciation was highly desirable, as it yielded an extremely useful theoretical framework. We are referring to what has come to be called the theory of *plate tectonics*, which was pioneered by D. P. McKenzie and R. L. Parker of Cambridge University, W. J. Morgan of Princeton, and X. Le Pichon of Lamont Geological Observatory.

The concept arises quite naturally when one leans back a little and looks at the motions on the surface of the earth that we have been describing in this chapter. We see great *rigid* slabs of ocean floor being created at ridges and consumed at trenches. Since not all the oceans are spreading at the same speed, we should expect to see, and do see, fault contacts between some of these great moving slabs as they grind past each other. And this is where we now come to the conceptual difference between the early drift pioneers and the modern analysts. Where Wegener and the others concentrated very heavily on the continents and their motions and regarded the ocean floor as something to be drifted through or across, the modern pioneers of plate tectonics pointed out that we should look at the earth's surface and its motions quite differently. Instead of subdividing the earth's crust into important continents and relatively insignificant ocean floors, we should rather dissect the earth's surface into a set of moving rigid slabs of rock which may be continents, ocean floors, or coupled

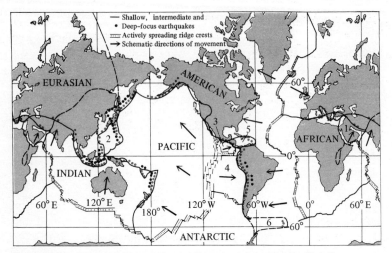

Figure 6-22 The division of the earth's crust into plates. The 6 major plates are named. 6 minor plates are numbered: (1) Arabian; (2) Philippine; (3) Cocos; (4) Nasca; (5) Caribbean; (6) Scotia. (*After F. J. Vine.*)

continent and ocean floor. These great slabs, of the order of 150 km thick, are usually called *plates*, and it is a fundamental tenet of plate tectonics that the earth's surface comprises a mosaic of such rigid plates which are in motion relative to each other and that it has been the continued interactions among these plates which has yielded most of the observed structural features of the crust. What then are the natural boundaries of these plates, if they are not to be merely the coastlines or the 500-fathom lines of the old pioneers? From what we have just said in this section, it should now be clear that there are three totally distinct kinds of boundary between plates—oceanic ridges, oceanic trenches, and the large faults, usually called *transform faults*. The ridge boundary is one separating two plates which are moving apart. It is a *constructive* boundary in the sense that *new plate*, that is, new ocean floor, is being created there, according to the ocean-floor-spreading concept. The oceanic trench boundary is the line along which one plate (which must be oceanic at this point since continents are not absorbed into the mantle) descends under another to be resorbed into the mantle. A trench boundary, often called a *subduction zone*, is therefore a *destructive* boundary, since the oceanic portions of a plate are destroyed there by plunging down into the mantle. The third boundary, the transform fault, is conservative and is neither constructive nor destructive, being merely the plane along which two adjacent plates slide past each other. In Fig. 6-22 we show the outlines of the plates of the earth's crust as they are defined by these three types of boundaries. We see there are six major plates and numerous smaller ones, but the theory is so new that the final outlines of the smaller slabs have not been unanimously agreed on.

Plate Motions and The Distribution of Earthquakes

On this concept the distribution of earthquakes about the earth's surface falls readily into account. The vast majority of earthquakes occur along plate boundaries because of the relative motion between the two plates. Most shallow mid-oceanic earthquakes are due to the jerky relative motions of two plates sliding past each other along transform faults (as in Fig. 6-19b). The circum-Pacific belt of earthquakes predominantly lies along various trench boundaries between plates, and in this case the earthquake shocks are a result of the downthrusting of old oceanic plate into the mantle beneath the other plate (see Fig. 6-23). The dipping plane along which the intermediate and deep focus earthquakes were seen to lie (Chap. 2) corresponds to the great tongue of cool ocean floor protruding into the warmer mantle. These earthquakes may be produced either by slip between the tongue and the mantle or by release of stresses in the cool slab as it adjusts to its new warm environment. The fact, also noted in Chap. 2, that no earthquakes occur at greater depths than about 700 km, is now readily explained by supposing that the descending plate has heated up to the temperature of its surrounding mantle by the time it has reached this depth and has been effectively resorbed in the earth's interior.

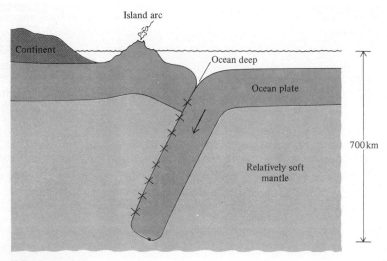

X–Earthquake foci

Figure 6-23 The earthquake activity around the rim of the Pacific is due in most areas to the downthrusting of the cool, rigid oceanic plate into the warm mantle.

Mountains and Plate Tectonics

Now we can see how the formation of mountain chains fits into the theory of plate tectonics. We saw in Chap. 4 that the cycle of events culminating in the formation of a folded mountain chain (e.g., the Andes) always begins with the accumulation of a long, linear prism of sediments known as a *geosyncline.* There we outlined the ensuing sequence of mechanical and thermal events which led to a lofty mountain belt. We ask therefore, now, how and where are such geosynclines formed, and equally important, how are they subsequently deformed and metamorphosed into mountain ranges?

We begin our answers to these questions by noting an important feature of the structure of geosynclines that has been known for many years: The detailed geological analysis of existing mountain chains had shown that a geosyncline really comprised two parallel linear features known as a *miogeosyncline* and a *eugeosyncline.* The miogeosyncline was composed of sediments which are characteristically deposited in shallow water. In sharp contrast the eugeosynclinal sediments were deep-water type. Now when we search the earth's surface for any possible present-day incipient geosynclines, unquestionably the most likely candidates are found along the coastlines of continents bounding oceans like the Atlantic, which are expanding by the ocean-floor-spreading mechanism. As an illustration we use the Eastern seaboard of North America. There we find a great sedimentary prism about 2,000 km long and about 250 km wide. It is located immediately to the seaward side of the continental slope, and seismic studies show it to reach a maximum thickness of about 10 km (see Fig. 6-24). This then is a growing *eugeosyncline*. Its associated miogeosyncline lies immediately on its

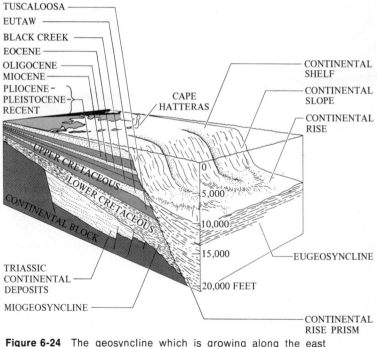

Figure 6-24 The geosyncline which is growing along the east coast of North America. (*After R. S. Dietz.*)

landward side, as we see in Fig. 6-24. It is the 5 km or so of shallow-water-deposited sedimentary rocks which underlie the continental shelf.

Such continental coastal couplets are envisaged to form in plate-tectonic theory as follows. A continent is imagined to split into two fragments which drift apart as new ocean floor is created between them along a central ridge (see Fig. 6-25a). The natural processes of erosion now begin the buildup of the geosynclinal couplet along the trailing edges of the two continents. Shallow-water sediments are laid down along the continental shelf to construct the miogeosyncline. Sedimentary material carried along undersea canyons across the continental slope spills out in great fans in much deeper water and begins the buildup of the eugeosyncline. The continental margin, which had been raised during the initial rifting, slowly sinks, thereby allowing as much as 5 km of miogeosynclinal sediment to be deposited in water which is much shallower than this. The sheer weight of this sedimentary load at the continental edge also causes sinking, as we would expect from our discussions of isostasy. After perhaps 100 million years we would have a well-developed geosyncline. As yet it is essentially undistorted and we are a long way from having a mountain chain.

The next stage in the mountain-building cycle is considered to be a breaking of the oceanic part of the plate, as shown in Fig. 6-25b, with the subsequent formation of a trench-type boundary or subduction zone just to the seaward side of the geosyncline. The ocean floor, instead of being coupled to and moving along with the continent, now begins to be thrust under the continent. This causes the *eugeosyncline* to collapse and be folded into severe contortions. Meanwhile, as the downthrusting ocean floor reaches a depth of about 100 km, the surrounding temperature is sufficiently high that the descending slab begins to melt partially. The resultant molten rock (*magma*) rises to the surface to erupt in lava flows. This high-temperature stage is one of general upwelling of molten rock into the higher crustal regions, and this is accompanied by thermal metamorphism of many of the eugeosynclinal rocks. The final stage of this *igneous* activity is the emplacement of granitic bodies into the upper crustal area. This

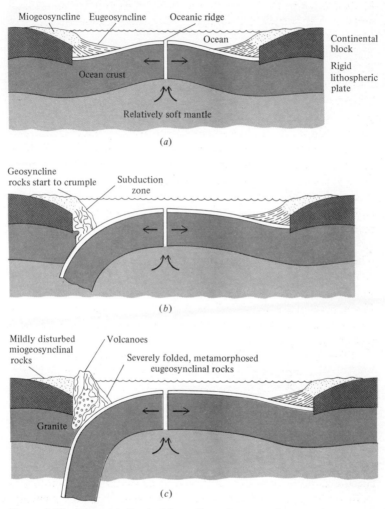

Figure 6-25 (a) to (c) Gradual formation of a coastal mountain range such as the Andes or the North American Cordillera. (*After R. S. Dietz, J. F. Dewey, and J. M. Bird.*)

sequence of events is illustrated step by step in Fig. 6-25*a* to *c*. In contrast to this violent mechanical and thermal history suffered by the eugeosynclinal sediments, the miogeosynclinal belt leads a

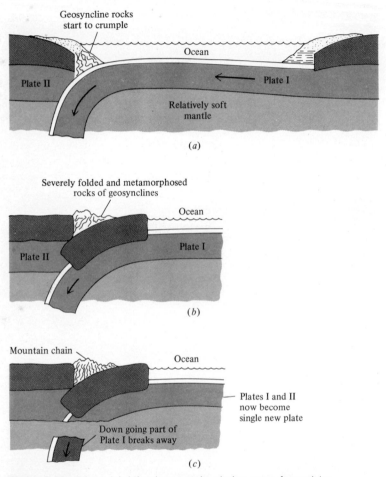

Figure 6-26 (*a*) to (*c*) Inland mountain chains were formed by collisions of this type between continents. The Himalayas, for example, were formed in this fashion when India drifted thousands of miles northwards from its position in Fig. 6-6 and ran into Asia. (*After R. S. Dietz, J. F. Dewey, and J. M. Bird.*)

much more sheltered life. These shallow-water deposits were laid down on top of hard continental rocks, and this strong basement affords them considerable protection during the mountain-building hiatus we have just described. They consequently suffer much less severe folding and metamorphism than do the eugeo-synclinal beds. Finally, as this great distorted geosynclinal prism cools and becomes more rigid, a general uplift takes place as the earth tries to return to isostatic equilibrium and allow this long chain to float at its proper level.

It seems likely that a history such as we have just described accounts for the origins of great coastal mountain belts like the South American Andes and the North American Cordilleras. But, you may object, not all mountain chains lie along the edges of continents. The Himalayas and the Urals, for example, are found deep within continental Asia. How does one account for them on plate-tectonic theory? The answer is quite simple and, in fact, was known to the early pioneers of continental drift. These mountains, lying well within continents, are the results of colli-sions between continents. In Fig. 6-26*a* to *c* we illustrate an idealized sequence of events involved in such a collision; Fig. 6-26*a* shows plate I underthrusting plate II along a trench boundary at the edge of plate II. We imagine that a geosyncline exists along the continental edges of both plates. As underthrust-ing proceeds, folding and metamorphism of the geosyncline on plate II will take place exactly as described above for the Andes. However, the continent on plate I is being brought ever closer to the crumpled geosyncline of plate II. Eventually the two con-tinents collide, with much folding of the geosynclinal rocks on both continental edges. Fragments of ocean floor are also caught up among the continental rocks. The bouyancy of the light continental rocks of plate I prevents it from being dragged down below plate II into the mantle, and in fact the motion of plate I relative to plate II is thus braked to a halt. The downhanging portion of plate I now probably breaks off and sinks deeper into the mantle until it is melted and reabsorbed there. Two originally distinct continents will in this way have been sealed together. The inland mountain chain therefore is a line of suture.

DRIVING MECHANISM

With our discussion of the origin and location of mountains according to plate-tectonic theory, we have come the full circle. We saw at the beginning of this chapter that Taylor's pioneering paper on continental drift in 1908 was entitled "Bearing of the Tertiary Mountain Belt on the Origin of the Earth's Plan." But in reviewing our comments on Taylor's work we are brought face to face again with the problem of what causes the plate motions. Taylor's dramatic concept that the earth had captured the moon in a close orbit about 100 m.y. ago and that this event had triggered drift is not accepted today. It is now generally considered that plates had been in motion for many hundreds of millions of years before the last episode that so intrigued Taylor, Baker, and Wegener, and it seems only reasonable to search for the cause of this behaviour within the earth itself. Furthermore, age determinations of the type described in Chap. 4, carried out on lunar rocks brought back on the Apollo missions, show that the moon has enjoyed a remarkably peaceful existence for at least 3 billion years.

When we look to the earth itself to provide the motive force for plate motion, the most obvious and probably the correct candidate is convection in the mantle as, we recall, Holmes and Hess believed. While on a short time scale the mantle may be regarded as a solid since it transmits seismic S waves, on the time scale of thousands and millions of years it behaves as a fluid, as

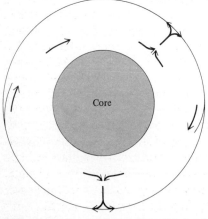

Figure 6-27 Convective motions driving the plates may involve the whole mantle, as shown in this simplified diagram. The arrows showing possible motions deep in the mantle are speculative.

we saw in Chap. 3. The pattern of the mantle convection cells, however, that is, their shape, dimensions, and locations, is not yet known. Some clues are provided by the motions and dimensions of the surface plates, but because of the considerable changes in the physical state of the earth with increasing depth, surface observations cannot be easily extrapolated downwards. Obviously, there must be upwelling of material beneath ocean ridges, horizontal motions just beneath plates, and downward motions near downgoing plates at subduction zones. A grossly simplified illustration of such possible convection motions is shown in Fig. 6-27. It must be remembered, however, that only those parts of the diagram pertaining to the outer part of the earth would be accepted by most scientists. That is, all are agreed that the moving plates are 100 to 150 km thick and that their *lower* boundaries are formed by the *upper* boundaries of the low-velocity zone. But here the agreement ends. Some confine the rest of the motion in the mantle to the low-velocity zone—that is, they believe convection is restricted essentially to the outer 400 km or so of the earth. Others, in line with Fig. 6-27, believe that the convection cells involve the whole mantle. Since there seems to be no way of actually *seeing* the cells, their outline must in the end be found by calculation. Efforts in this direction have so far been unsuccessful, mainly because the variation with depth of the *viscosity* of the mantle is not well known. Viscosity is the property which describes how easily a fluid will flow. Thus water has a low viscosity, syrup is much more viscous, while pitch, at ordinary room temperatures, is far more viscous than both. When pitch is heated, however, its viscosity drops considerably and it flows readily. In the same way, the variation of temperature in the mantle will cause significant variation in the mantle viscosity. The mantle temperature profile (Fig. 6-28) is not well known, however, and this uncertainty is therefore carried over into the viscosity values. A further complication arises when we consider the heat sources which drive convection. The usual elementary situation in convection problems envisages a mass of fluid heated from below and cooling at the top. In the case of the earth, the radioactive elements K, U, and Th must provide much of the heat, yet their distribution is poorly understood. While the

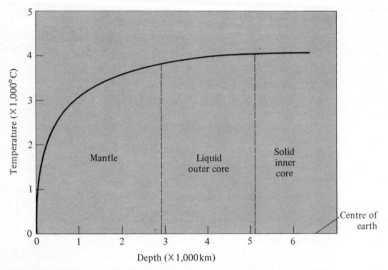

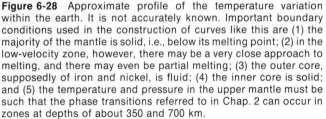

Figure 6-28 Approximate profile of the temperature variation within the earth. It is not accurately known. Important boundary conditions used in the construction of curves like this are (1) the majority of the mantle is solid, i.e., below its melting point; (2) in the low-velocity zone, however, there may be a very close approach to melting, and there may even be partial melting; (3) the outer core, supposedly of iron and nickel, is fluid; (4) the inner core is solid; and (5) the temperature and pressure in the upper mantle must be such that the phase transitions referred to in Chap. 2 can occur in zones at depths of about 350 and 700 km.

continental crustal rocks harbour much radioactivity, the amount and location in the mantle are unclear. These factors and others have accordingly so far prevented the satisfactory calculation of the details of the convective motions at depth.

FINALE

We have now seen some of the remarkable features of the earth's properties and behaviour that have been discovered in the twentieth century. While some broad outlines and some important details had been sketched by 1950, an extraordinary advance has occurred since then. The 1950s was a time of honing new techniques and of producing new information with these. The 1960s and early 1970s have been a period of interpretation and great syntheses. They have encompassed a decade of revolution which has made the study of the earth probably the most exciting field in science today.

Bibliography

The following list should help the reader pursue further the ideas discussed in "Planet Earth."

Bolt, B. A., The Fine Structure of the Earth's Interior, *Scientific American*, vol. 228, pp. 24–33, March 1973.

Bott, M. H. P., "The Interior of the Earth," Edward Arnold, 1971.

Cailleux, André, "Anatomy of the Earth," McGraw-Hill, 1968.

"Continents Adrift—Readings from Scientific American," Freeman, 1972.

Faul, H., "Ages of Rocks, Planets, and Stars," McGraw-Hill, 1966.

Gass, I. G., Smith, P. J., and Wilson, R. C. L., "Understanding the Earth," Open University Press, 1971.

Hodgson, J. H., "Earthquakes and Earth Structure," Prentice-Hall, 1964.

Holmes, Arthur, "Principles of Physical Geology," Nelson, 1965.

Jacobs, J. A., Russell, R. D., and Wilson, J. T., "Physics and Geology," 2d ed., McGraw-Hill, 1974.

King-Hele, D., "Observing Earth Satellites," Macmillan, 1966.

Mason, B. H., "Meteorites," Wiley, 1962.

Payne-Gaposchkin, C., and Haramundanis, K., "Introduction to Astronomy," Prentice-Hall, 1970.

Richter, C. F., "Elementary Seismology," Freeman, 1958.

Strangway, D. W., "History of the Earth's Magnetic Field," McGraw-Hill, 1970.

Wegener, A., "The Origin of Continents and Oceans," Dover, 1966.

York, D., and Farquhar, R. M., "The Earth's Age and Geochronology," Pergamon, 1972.

Index

Index